When the Universe Knocks at Your Door

How Planets, Stars and Other Cosmic Objects Affect Our Daily Lives

By

Irina Mullins

Digital Green World Publishing

When the Universe Knocks at Your Door: How Planets, Stars and Other Cosmic Objects Affect Our Daily Lives

ISBN: 978-1-936496-01-3

For contact information, please visit our website,

www.digitalgreenworld.com

To contact the author of this book, please visit the *Science Today and Tomorrow* website:

www.sciencetodayandtomorrow.com

Table of Contents

Preface

Our Universe is full of contrasts. It is the place where the hostile environment of interplanetary and interstellar space exists along with livable planets. It is the place where massive stars end their lives in powerful destructive explosions, giving away building material for new stars, planets and any life forms that might emerge on those planets.

When we ponder about the ways the Universe can affect our lives, we usually think of such cataclysmic events as blasts from supernovae or asteroid impacts. However, the Universe "touches" our planet in less dramatic ways on a daily basis. In fact, it reaches us every day in so many different ways that it would be difficult to discuss all of them in one book.

Written for the general public, this book answers questions about some interesting celestial objects and cosmic events that affect our lives and our planet on a regular and not-so-regular basis.

Introduction and a Very Brief Review of the Solar System and the Universe

More than two thousand years ago, the Greek philosopher Aristotle proposed that our Universe was made of five elements. The first four elements were earth, water, air, and fire. According to Aristotle, the Sun, the Moon, the other planets, and all the stars were made of the fifth element, called the "ether" or "quintessence." The fifth element was believed to be a "perfect" material, and all the cosmic objects composed of it were believed to be perfect and unchanging.

Today, we know that the Earth and the Sun, planets and their moons, stars and galaxies belong to the same realm of the constantly changing physical world. In this world, the Earth is not alienated from the rest of the Universe. In fact, the Earth and all the living beings on our planet experience the presence and the influence of the Universe every day and every moment.

Our planet belongs to the planetary system that is called the Solar System. When people think about the Solar System, they usually picture in their mind the Sun and the eight major planets (Mercury, Venus, Earth,

Mars, Jupiter, Saturn, Uranus and Neptune) that orbit the Sun. However, the Solar System is actually a rather "crowded" place. It consists of the Sun and a huge number of cosmic objects, including eight major planets and their moons, gas and dust, millions of asteroids, billions of comets and other icy objects. All these "creatures" great and small have the ability to affect our planet and our lives.

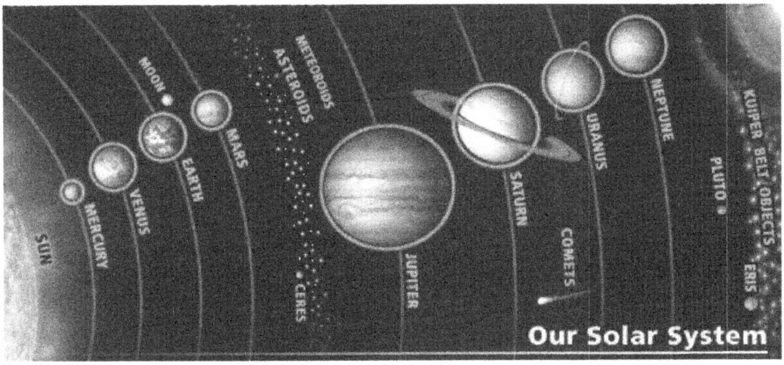

Image I.1. The Solar System. The image is not to scale. Image Credit: NASA/JPL.

The Sun is the central and only star of our Solar System. The Sun's mass makes up 99.8 percent of the mass of the entire Solar System. The Sun's gravity holds all the objects in our Solar System and makes them orbit the Sun. Almost all the planets, except for Mercury and Venus, are orbited by their moons. Today, astronomers know of more than 140 moons that orbit the planets and follow the planets in their journey around the Sun.

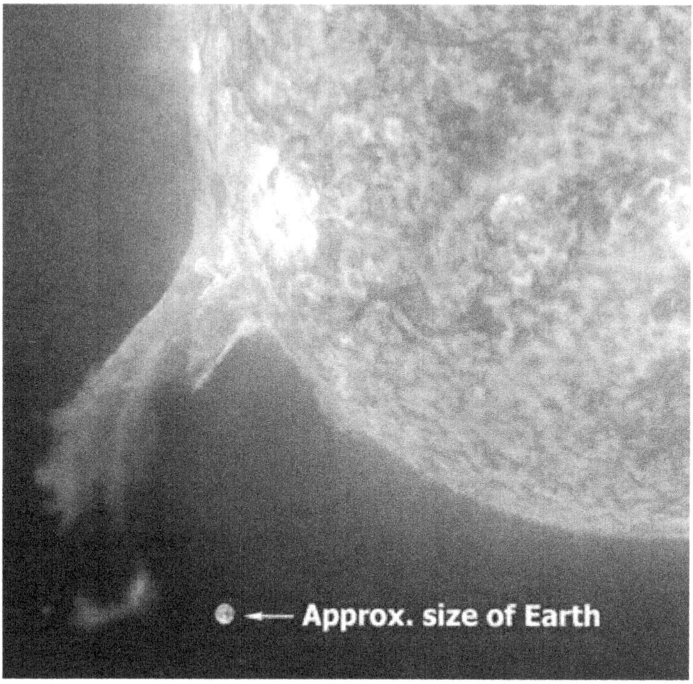

Image I.2. The Earth and the Sun. This illustration shows the approximate size of the Earth compared to the Sun. The giant coronal mass ejection in this image blasted off the Sun in October 2003. The image was taken by the SOHO spacecraft. Image Credit: NASA and ESA.

On the Earth, we conventionally measure distances in miles or kilometers; however, these units of measurement are not large enough when it comes to the Solar System. For example, the average distance from the Earth to the Sun is 149 million kilometers (93 million miles). Measuring distances between planets and the Sun in kilometers or miles is like measuring distances between cities on our planet in centimeters or inches.

This is why astronomers use the astronomical unit (AU) to measure distances in the Solar System or within any other planetary system. One astronomical unit equals the average distance from the Earth to the Sun. For example, Neptune, the farthest planet from the Sun, is about 30 times farther away from the Sun than the Earth. Therefore, the average distance between the Sun and Neptune is about 30 astronomical units.

Many comets that orbit the Sun are located much farther from the Sun than Neptune. Comets are small, irregularly shaped objects composed mostly of a mixture of water ice, dust, carbon- and silicon-based compounds. They populate two regions of the Solar System: the Kuiper Belt and the Oort Cloud. Located beyond the orbit of Neptune, the Kuiper Belt contains at least 200 million icy objects that have not changed much since the Solar System formed 4.6 billion years ago. The Oort Cloud extends at a distance of about 50,000 astronomical units! Astronomers estimate that the Oort Cloud contains billions of comets and other icy objects.

Whereas the majority of comets permanently reside in the Kuiper Belt and the Oort Cloud, some comets leave these regions of the Solar System and travel along very elongated orbits toward the Sun.

Image I.3. Comet C/2002 V1 (NEAT). True-color image of this comet was obtained on January 29, 2003. Image Credit: European Space Agency.

When a comet approaches the Sun, the solar radiation heats the comet. The comet's ice transforms into a vapor and releases dust particles. The Sun's radiation and the solar wind, which is a stream of very fast-moving charged particles blowing from the Sun, sweep the evaporated cometary material so that it forms a long tail.

Astronomers classify comets in accordance with their orbital period, which is the time it takes an object to complete one trip around the Sun. Short-period comets have orbital periods of less than 200 years. For example, Comet Halley's orbital period is 76 years. Short-period comets most likely come from the Kuiper Belt. They were probably "pushed" into the orbits taking them close to the Sun by gravitational interaction with other cosmic bodies.

Long-period comets most likely come from the Oort Cloud. Their orbital periods are more than 200 years. Typically, long-period comets have their orbital periods of thousands of years. For example, Comet Hale-Bopp completes one trip around the Sun in about 4,000 years!

Asteroids represent another "family" of cosmic objects populating the Solar System. They are basically pieces of rock of irregular shape (sort of like potatoes). Some of them are covered with frozen water. Millions of asteroids reside in the main asteroid belt, which is the region of space between the orbits of Mars and Jupiter. Scientists propose that the asteroids in the main asteroid belt never had a chance to form a planet because Jupiter's gravity kept pulling them away from each other.

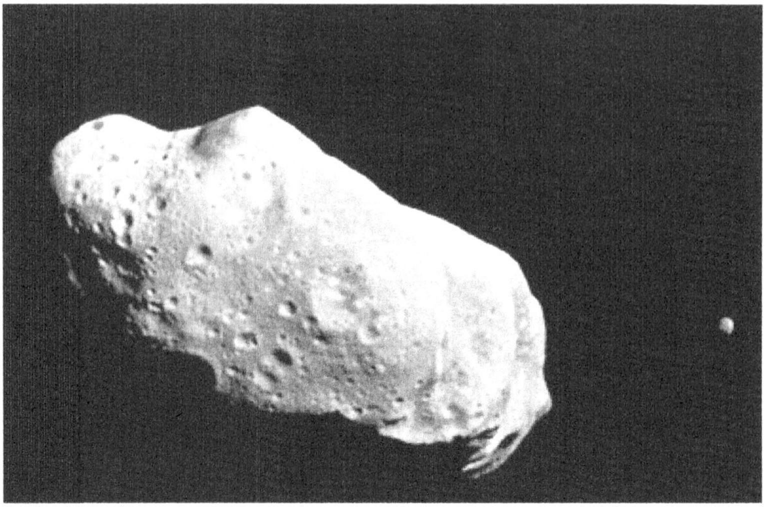

Image I.4. Asteroid 243 Ida and its moon Dactyl. The image was transmitted to Earth from the NASA's Galileo spacecraft. Image Credit: NASA.

Besides asteroids, the main asteroid belt is also populated with meteoroids. The only difference between asteroids and meteoroids is their size. Meteoroids range in size from a grain of dust (micrometeoroids) to about 100 meters across. Any space rocks that are large than that are called asteroids.

Similarly to comets, some asteroids have more elongated orbits that periodically bring them close to the Sun. As they travel in space, they may also cross the orbit of the Earth. Sometimes such comets and asteroids (or meteoroids) collide with the Earth and the other planets. Asteroids or meteoroids that fall to the Earth's surface are called meteorites.

Meteorites vary in size. The majority of meteorites are the dust-sized particles (micrometeorites) from space that fall to our planet. Some meteorites can be as big as boulders, and a few meteorites can be real giants. The largest meteorite ever found on the Earth is the Hoba meteorite. Discovered in Namibia in 1920, it weighs approximately 54 tons (119,000 pounds).

Meteorites have a burned exterior because air friction heats them to high temperatures and melts them (completely or partially) before they reach the Earth's surface. Meteorites move very fast, and they crash into the Earth with tremendous force. When large meteorites strike the Earth, they create impact craters, which are circular craters with a raised rim.

The Earth's geological activity continuously changes its surface, erasing ancient impact craters. Nevertheless, scientists can still find some very old impact craters on the Earth. So far, more than a hundred ancient impact craters have been found on our planet.

There are eight major planets in the Solar System: Mercury, Venus, Earth, Mars, Jupiter, Saturn, Uranus and Neptune. Mercury, Venus, Earth and Mars are called *terrestrial planets* because in some ways they are similar to the Earth (although they also differ from each other in many of their properties).

The terrestrial planets are the planets closest to the Sun. They are relatively small in size. Each terrestrial planet has a solid rocky surface. Of all the major planets, Mercury has the thinnest atmosphere. Venus has a very thick atmosphere composed mostly of carbon dioxide. Thanks to its atmosphere, Venus is famous for its runaway greenhouse effect. A very thin atmosphere made mostly of carbon dioxide covers Mars. The atmosphere of the Earth is 77 percent nitrogen, 21 percent oxygen, and 1 percent argon. It also has some variable amounts of water vapor, and trace amounts of carbon dioxide and other gases.

Jupiter, Saturn, Uranus and Neptune are called *Jovian planets* because their composition is similar to that of Jupiter. Jovian planets are also called giant planets because of their large size and mass. They do not have solid surfaces. Some amounts of rocky materials are incorporated deep in the cores of Jovian planets. Jupiter and Saturn contain huge amounts of hydrogen and helium. Uranus and Neptune contain significant amount of frozen water, ammonia, methane and carbon monoxide.

The Solar System belongs to the Milky Way Galaxy, a huge collection of stars in an interstellar medium (clouds of gas and dust). On a dark night, we can see the disk and the bulge of the Milky Way Galaxy as a milky band of light stretching across the sky, which is why we call it the Milky Way. If we look at it with binoculars or a telescope, we can see more individual stars in it.

All the individual stars that we can see in the night sky without a telescope belong to the Milky Way Galaxy. Actually, we can also see a few other galaxies without using a telescope. For example, the Large and Small Magellanic Clouds are easily observed from the southern hemisphere. On dark and moonless nights, we can see the nearby Andromeda Galaxy, which is also called M31, with the naked eye. However, we cannot discern individual stars in those galaxies.

Very rarely, we can see individual stars in the nearby galaxies without using a telescope when these stars go supernovae. For example, a supernova called SN 1987A was visible to the naked eye in the night sky in 1987. It occurred in the Large Magellanic Cloud.

The Sun is one of more than 300 billion stars populating the Milky Way Galaxy. Proxima Centauri, the star nearest to the Sun, is located about 270 thousand astronomical units away from us. That's a huge number! Obviously, the astronomical unit is not large enough to measure distances between stars or galaxies. Therefore, astronomers measure these distances in light years.

A light year is the distance light travels in one year (remember that light travels at a velocity of about 300,000 kilometers per second, which is the same as about 186,411 miles per second). One light year equals 9.4 trillion kilometers (about 6 trillion miles.) For example, the Sun is located at a distance of about 28,000 light years from the center of the Milky Way Galaxy. In empty space, a ray of light would travel such a distance in 28,000 years!

Image I.5. The Milky Way Galaxy. This is a picture of the real Milky Way taken by the satellite COBE. Image Credit: The COBE Project, DIRBE, NASA.

The Milky Way Galaxy is a spiral galaxy, and the Solar System is located in one of the spiral arms of the galaxy. The disk of our galaxy is so big that a ray of light would have to travel about 100,000 years in order to cross the galactic disk from one side to the other!

The Milky Way Galaxy is not the only galaxy in our Universe. Astronomers estimate that the observable Universe is populated with more than 3 billion galaxies.

Image I.6. Hubble Space Telescope took this image of galaxies in 2003. Galaxies are huge collections of stars. Image Credit: NASA, ESA, J. Blakeslee and H. Ford (Johns Hopkins University).

In recent decades, astronomers have discovered that planets, stars, galaxies and all the visible matter of the Universe make up only about 5 percent of the Universe. The other 95 percent is made up of dark matter and dark energy. Scientists still do not have a good understanding of the nature of dark matter and dark energy.

Part I. Comets and Asteroids

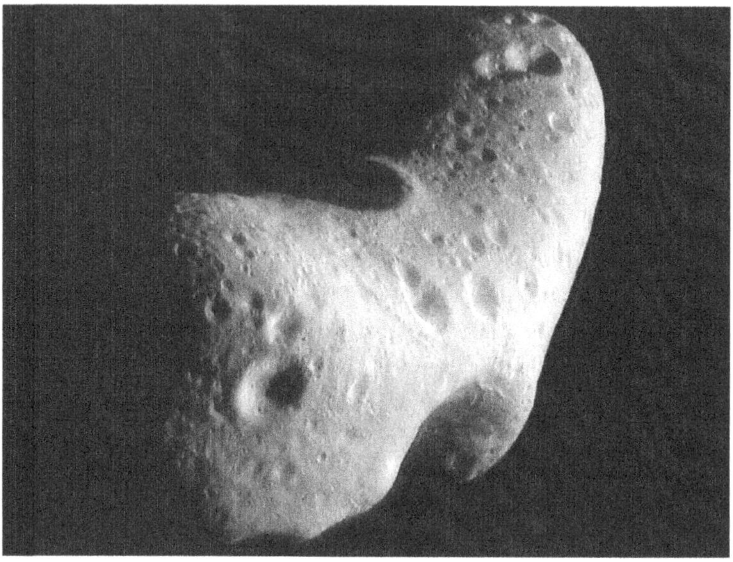

In 2000, NASA's Near Earth Asteroid Rendezvous mission took this image of Eros, an asteroid with an orbit that periodicaly brings it somewhat close to Earth. Image Credit: NASA/JHUAPL.

1.1. How do shooting stars change our planet?

As the Earth orbits the Sun, it travels about 108 thousand kilometers (67 thousand miles) in one hour. This means that every day the Earth travels approximately 2.6 million kilometers (1.6 million miles) in its orbit!

The Earth moves through space at a nearly constant speed. We do not feel its motion because we can feel motion only when the motion changes. For example, when you travel by airplane, you feel pressed back into the seat when the airplane accelerates in order to take off. You find yourself pushed forward when the airplane slows down during its landing. However, when the airplane is up in the air and flying at a constant speed, you don't "feel" its motion.

Of course, in the real world, turbulence can remind you that the airplane is moving in the air. But we do not have to worry about turbulence when it comes to the Earth's motion in space. The Earth's "ride" around the Sun is very smooth because the space between planets in the Solar System is almost empty. However, it is not completely empty. It is actually filled with interplanetary gas and dust. Scientists estimate that if we could somehow collect all the interplanetary matter found in space within the orbit of Jupiter, it would be sufficient to "build" a moon as big as 25 kilometers (about 16 miles) in diameter!

Shooting stars often remind us that interplanetary space is not a void. Shooting stars look like stars falling towards us from the night sky, but they have nothing in common with real stars. Shooting stars are streaks of light produced when interplanetary dust particles burn up during their entrance into the Earth's atmosphere. (The official name for shooting stars is meteors.) When these particles enter the atmosphere, they typically move almost 100 times faster than a rifle bullet!

Interplanetary dust particles are very small (typically less than a few hundred micrometers across) and very light (their mass is usually about one millionth of a gram). Yet, they can look spectacular and star-like when they enter the Earth's atmosphere. Every day many trillions of these particles find their way from outer space to the Earth's surface. So, when you wash your car next time or when you do some dusting at home, remember that you are wiping off a certain amount of cosmic dust mixed up with Earth-bound dust.

You can actually collect a small sample of these cosmic particles yourself. All you need is a magnet and a large plastic bag or plastic sheet. You will have to place the bag or sheet on the ground somewhere outdoors and wait for a few hours. Then, you will have to run a magnet just above the plastic bag or sheet and look closely at the magnet. Because some of the cosmic particles include metals, you will find specks of cosmic dust that the magnet has picked up from the plastic bag or sheet.

Particles that enter the Earth's atmosphere move very fast, experiencing strong friction against the air they pass through. The friction increases their temperature to more than 1,000 degrees Celsius (more than 1,832 degrees Fahrenheit). Particles that are larger than 2 millimeters in diameter are heated so much that they vaporize at about 60 kilometers (37 miles) above the Earth's surface, producing a meteor. Smaller particles can survive their trip through our atmosphere because they are heated less. In the end, they settle on the Earth's surface.

Scientists estimate that about 40,000 tons (80,000,000 pounds) of cosmic dust land on the Earth's surface every year. The amount of "incoming" cosmic dust can be estimated or measured in a few different ways: from the rate of accumulation of rare elements, such as iridium and osmium, in polar ice and in deep-sea sediments (these rare elements are thought to be linked to cosmic dust); from meteor radar observations, remote laser sensing and from measurements made by high-altitude aircraft.

Some meteors are random events; however, many observed meteors are part of meteor showers that occur at the same time every year, originating from the same region of the sky. This tells us that meteors from the same meteor shower should have the same origin. Indeed, they usually come from the same source—comets, which are chunks of frozen water with some dust embedded in the ice.

As a comet moves in its orbit around the Sun, it periodically gets closer to the Sun. When this happens, the solar heat increases the comet's temperature and the comet gives off gas and dust, leaving a trail of dust in its orbit. When the Earth travels through a stream of particles of dust left behind a decaying comet, the particles enter the Earth's atmosphere, and we see them as a meteor shower.

To visualize this, imagine a truck hauling sand. If the sand is not covered properly, the truck will be leaving behind itself a trail of sand. If you happen to drive behind that truck, some sand particles will end up hitting your car.

Watching meteor showers from the Earth can be fun; however, those fast-moving dust particles can be dangerous beyond the Earth's atmosphere. When they collide with a spacecraft or with a space-walking astronaut, a collision can make a hole in the spacecraft components or in a spacesuit. It can also create ions capable of short-circuiting electrical systems of the spacecraft. Engineers have to design special protective "guards" to shield a spacecraft or a space station, as well as astronauts, from the "super-fast" cosmic dust.

Comets are not the only objects that can "supply" our Solar System with gas and dust. Planets and moons can also produce dust by volcanism and outgassing. Planets and moons can even eject some dust into space when they are hit by asteroids and meteoroids. Cosmic dust is produced in collisions between asteroids as well.

Sometimes, cosmic dust can be created when a large meteoroid disintegrates in the upper layers of our atmosphere. For example, in September 2004, a house-sized meteoroid entered the Earth's atmosphere. Heated by air friction to a very high temperature, it ended its existence in a powerful explosion high above the Earth's surface. The amount of energy released during the explosion was equivalent to the amount of energy of the atomic bomb that was dropped on the Japanese city of Hiroshima. The explosion produced about 1,000 tons (about 2,000,000 pounds) of dust.

Cosmic dust particles do make our planet gain weight. They can also affect the Earth's atmosphere and climate. When they evaporate in the Earth's atmosphere, they release magnesium, iron and silica that might react with chemicals in the atmosphere, eroding the ozone layer in the stratosphere. They can also trigger the formation of the highest clouds in the polar regions of the Earth's atmosphere. This happens when drops of water stick to specks of cosmic dust.

Moreover, when present in our atmosphere, cosmic dust particles can cool the Earth by scattering the sunlight away from our planet, and they can warm the Earth by absorbing sunlight.

Cosmic dust particles can also affect life on the Earth in some unexpected ways. A few billion years ago, they might have played a role in the emergence of life on our planet, because they might have delivered some organic compounds to the Earth. Today, they continue acting as a special fertilizer. They bring iron, which is needed for photosynthesis, to our oceans. This so-called ocean iron fertilization can trigger plankton blooms.

Shooting stars can do many things to our planet. So, when you see a shooting star next time, remember that there is more to that shooting star than meets the eye.

1.2. How do meteorites make our planet more "precious?"

Comets, asteroids and meteoroids shower our planet with cosmic dust every day. Yet, on some occasions, the cosmos sends something very precious to the Earth. Something like diamonds. Alas, they do not fall from the sky on their own. They reach the surface of the Earth as part of meteorites.

Diamonds that are found inside meteorites are extremely tiny: a few trillions of such diamonds can be placed on the head of a pin. For example, in 1981, researchers from the Smithsonian used X-rays to study a large iron meteorite discovered in Antarctica. The X-ray image showed that the meteorite was punctured with microscopic diamonds. They may have formed in an immense collision between asteroids in the main asteroid belt.

Some diamonds may be created when meteorites crash into our planet. Such impacts are capable of producing the extreme heat and pressure required for the formation of diamonds. This theory agrees with the fact that diamonds have been found in impact craters. For example, scientists discovered diamond clusters in the Popigari Crater. Located in Siberia, this 35-mile-wide impact crater formed about 35 million years ago when a meteorite smashed into the Earth's surface. In the 1960s, scientists found microscopic diamonds in the remnants of the Canyon Diablo meteorite. This is the meteorite that created the famous Meteor Crater in Arizona.

Image 1.1. Microscopic diamonds may have formed when a meteorite struck the Earth and created Arizona's Meteor Crater. Image Credit: NASA.

On the Earth, traditional diamonds form over the course of billions of years. Their formation occurs deep beneath the surface of the planet, under conditions of intense heat and pressure. Yet, some space diamonds delivered to the Earth by meteorites formed in a very different way and much faster. They were created in a supernova explosion far away from the Solar System, and the presence of xenon in these diamonds gives us a hint about their origin.

The diamond creation process in supernovae is strikingly different from that inside the Earth. In supernovae, the process lasts for a fraction of a second, incorporating xenon gas into these "supernova" diamonds. In 1987, scientists from the University of Chicago discovered a meteorite with "supernova" diamonds that turned out to be older than our Solar System!

Diamonds that "come with meteorites" are so tiny that they would never draw the attention of any jeweler. However, Jupiter and Saturn might be able to boast of some real diamond rains as well as big diamonds. Of course, scientists have never had a chance to walk in Saturn's diamond rain. Nevertheless, two planetary scientists, Mona Delitsky from California Specialty Engineering in Flintridge and Kevin Baines from the University of Wisconsin-Madison, are fairly certain that diamond rain is real.

They propose that lightning storms in the upper atmospheres of Jupiter and Saturn transform methane into carbon soot. When the soot slowly descends through gaseous and liquid hydrogen in the direction of the planets' rocky cores, it is subjected to increasing pressure and temperature that transform the soot into graphite. Then, the graphite becomes compressed into solid diamonds that keep falling toward the planet's core.

When the solid diamonds reach the region of the planet's interior with a temperature of about 8,000 degrees Celsius (14,432 degrees Fahrenheit), they melt and form liquid diamond raindrops.

Delitsky and Baines estimate that Saturn's interior might contain about 10 million tons (about 20 billion pounds) of diamonds! The majority of these diamonds are no bigger than 1 millimeter (0.04 inch) in diameter, but some diamonds might measure up to 10 centimeters (4 inches) across!

Not all scientists agree with Delitsky and Baines. However, Jupiter and Saturn are not the only planets with extreme conditions in their interiors. Marvin Ross and his colleagues from Lawrence Livermore National Laboratory propose that Uranus and Neptune might have interior conditions suitable for the formation of diamonds.

With all those space diamonds from stars and planets, meteorites and micrometeorites, it looks like the Universe is richer than one might think, and it keeps sharing its diamond reserves with our planet.

1.3. Can another "Chelyabinsk" meteorite strike your city or town?

There are no traffic lights in our Solar System, and the "Chelyabinsk" meteorite, the largest space rock that has hit the Earth for the last hundred years, just recently reminded us about this.

A little after 9 a.m. (local time) on February 15, 2013, the 20-meter-wide (92-feet-wide) space rock flew as a glowing fireball over the city of Chelyabinsk in Russia. It was later named the Chelyabinsk meteorite. Travelling at a speed of about 68,400 kilometers per hour (43,200 miles per hour), the space rock blew apart 45 kilometers (18.5 miles) above the city of Chelyabinsk, reaching the ground as numerous meteorites. Its explosion was the largest space-rock airburst since the Tunguska explosion in 1908!

At the moment of the explosion, the Chelyabinsk meteorite looked 30 times brighter than the Sun, causing severe skin and retinal burns on people located a few miles from the meteorite. The total weight of the recovered fragments of the exploded meteorite was approximately 4 to 6 tons (8,000 to 12,000 pounds), which was only a fraction of the original space rock. The largest single piece weighed about 650 kilograms (1,433 pounds). It plunged into Lake Chebarkul, located

70 kilometers (44 miles) west of the city of Chelyabinsk, cutting an 8 meter-wide (26 feet-wide) hole in the ice that covered the lake. It was later recovered from the bottom of the lake.

When the space rock exploded above the city of Chelyabinsk, it released an amount of energy equivalent that in approximately 500 kilotons of TNT. In other words, the explosion of the Chelyabinsk meteorite released 33 times more energy than the explosion of the atomic bomb dropped by the US on the Japanese city of Hiroshima on August 6, 1945!

The meteorite left more than 1,200 people injured, mostly by flying glass and building debris. Within 88 kilometers (55 miles) on either side of the space rock's trail, its shockwave knocked people off their feet, shattered windows in about 3,600 apartment blocks, and made a factory roof collapse.

The violence of the Chelyabinsk meteorite's impact should be a warning message for humankind. There are no traffic lights in interplanetary space, so when the orbits of the Earth and any asteroid or comet cross, and both the Earth and an asteroid or a comet arrive at the "intersection" at the same time, there will be some sort of collision.

If the asteroid or comet is big enough, a collision between them can be devastating, creating impact craters on the surface of our planet and sometimes killing many living creatures or even causing mass extinctions on the Earth. The ancient impact craters found on the Earth remind us that such collisions happened in the past. About every million years, a 1.5-kilometer (almost 1-mile) asteroid or comet collides with the Earth. A collision between a kilometer-wide asteroid and the Earth can release energy that would be equivalent to approximately a million 1-megaton atomic bombs. It might result in the extinction of huge number of the species living on our planet.

About every 50 to 100 million years, the Earth collides with an asteroid as large as the asteroid that created the Chicxulub Crater. The Chicxulub Crater is one of the most famous and largest impact craters on the Earth. It measures about 10 kilometers (6 miles) across, and it is located in the Yucatan Peninsula, Mexico. About half of the Chicxulub Crater is submerged in the Gulf of Mexico, and the other half is located on land, remaining hidden beneath many meters of sediment.

Many scientists propose that when the meteorite that created the Chicxulub Crater smashed into the Earth, the impact changed the conditions on our planet very drastically, causing the extinction of dinosaurs and other life forms on the Earth about 65 million years ago.

Another famous impact crater is the Barringer Meteorite Crater, which is located in Arizona. It is 1 kilometer (0.6 miles) wide and approximately 230 meters (750 feet) deep. Scientists determined that this impact crater was created more than 50,000 years ago when a huge meteorite crashed into the Earth. The energy released during the impact was equivalent to 2.5 million tons of TNT.

These are the cautionary stories from the Earth's past. However, astronomical observations of comets, asteroids and meteoroids populating our Solar System remind us that devastating collisions will also happen in the future.

Where does the danger come from? The orbits of most asteroids lie between the orbits of Mars and Jupiter and they never cross the orbit of the Earth. However, a relatively small number of asteroids have orbits that do intersect the Earth's orbit. Astronomers call them *Earth-crossing asteroids* or *Apollo asteroids* (named after *Apollo*, the first discovered Earth-crossing asteroid).

A few hundred Earth-crossing asteroids are officially considered as potentially dangerous. They measure more than 150 meters (492 feet) in diameter, which is three times larger than the size of the meteorite that created the Barringer Meteorite Crater, and they move in orbits that bring them very close to the Earth. These asteroids will eventually collide with the Earth sometime in the future.

Comets also occasionally cross the orbit of the Earth, and they can collide with the Earth or any other planet just like asteroids can. Very recently, in 1994, astronomers observed how Comet Shoemaker-Levy 9 collided with the planet Jupiter. Before its collision with the planet, the comet felt apart into a few pieces. The largest piece was less than 1 kilometer (0.6 mile) across. Each piece of the comet smashed into the atmosphere of Jupiter, creating huge fireballs that measured hundreds of kilometers across. The largest fireball was even bigger than the Earth!

Meteoroids, asteroids and comets that have sizes ranging from meters to tens of kilometers and whose orbits come too close to the Earth's orbit are collectively known as Near-Earth objects (NEOs). So far, astronomers have discovered about 10,000 NEOs.

NASA scientists predict that no large asteroid or comet will strike the Earth in the next few centuries. But what about a not-so-large but still dangerous asteroid? Something similar to the Chelyabinsk meteorite, perhaps?

According to Bill Cooke, the head of NASA's meteoroid environment office at Marshall Space Flight Center, something similar to the Chelyabinsk meteorite might happen again in the near future. Peter Brown, a planetary scientist from the University of Western Ontario, agrees with Cooke. Brown estimates that an event similar to the Chelyabinsk meteorite should occur every 30 to 40 years.

When and where exactly will the next "Chelyabinsk-like" meteorite hit the Earth? Scientists do not have a good answer to this question. After all, we have to admit that the Chelyabinsk meteorite took us by complete surprise.

1.4. Where does our water come from?

It is nice to have a cup of coffee or tea in the morning. But where does that water in your coffee come from and where did all the water on the Earth come from? Surprisingly, not all the water on our planet is native to the Earth. Most of it must be of extraterrestrial origin, and the question about its origin takes us back in time, all the way to the period of formation of the Solar System.

Although scientists cannot travel back in time, they theorize that the Solar System formed from a disk of gas and dust about 4.5 billion years ago (theories and observations of young stars and their planetary systems support this idea).

The first 500 million years of the Earth's existence are called the Hadean Eon. We do not have much data for the Earth about that period because few rocks found on the Earth are older than 3.8 billion years (submersion in the oceans, plate tectonics, and erosion have been continuously "renewing" the Earth's surface since then). However, scientists propose that during the Hadean Eon, the Earth and its Moon, as well as the other planets, were colliding very frequently with comets and asteroids.

Image 1.2. The Center of the Orion Nebula that is located 1,500 light-years away from the Earth. The Nebula is filled with glowing protoplanetary disks which might eventually form planetary systems. Image Credit: NASA, C.R. O'Dell and S.K. Wong (Rice University).

Because the Earth and the Moon are very close neighbors, the Earth collided with comets and asteroids in the distant past just as frequently (probably even more frequently) as the Moon did. Therefore, the frequency of the "prehistoric" collisions can be estimated by counting ancient impact craters found on the Moon.

It follows from the studies of the Moon's impact craters that the young Earth experienced very frequent collisions with asteroids and comets of different sizes. However, collisions with very large asteroids were not frequent. During the Hadean Eon, the Earth was colliding with asteroids that were larger than 200 kilometers (120 miles) in diameter about once every 20 million years.

A collision with an asteroid larger than 100 kilometers (60 miles) in diameter could easily cover large areas of the Earth with molten rock. However, if the Earth had time to "chill out" a little between such catastrophic collisions, then some water that vaporized during one collision could rain down and remain on the Earth's surface until the next impact.

Can this theory be correct? Scientists have recently analyzed microscopic zircon crystals found in some minerals on the Earth. These crystals existed during the Hadean Eon, and their composition indicates that there might have been some amount of liquid water on the surface of the Earth at that time.

However, the intensive bombardment of our planet by comets and asteroids of different sizes during the Hadean Eon must have made the surface and the atmosphere of the young Earth so hot that most of the surface liquid water boiled and evaporated, escaping from the planet into space. This means that a significant fraction of the water that covers 71 percent of the Earth's surface today must have been delivered to our planet later, after the Earth cooled off.

So, how did water find its way to our planet later? According to one theory, water-rich asteroids and meteoroids brought water to the Earth when they were crashing into our planet. Another theory proposes that comets delivered water to our planet in a similar way.

To test these theories, scientists use the following interesting fact: for each 3,200 "regular" water molecules (H2O) in the Earth's oceans, there is one heavy water molecule. A heavy water molecule includes deuterium, a hydrogen atom with an extra neutron. Thanks to the extra neutron, a heavy water molecule is actually heavier than a "regular" water molecule.

Whatever cosmic object could have brought water to the Earth, it should have had the right amount of deuterium in its water. Scientists have confirmed that water-rich asteroids and some meteorites do have the same ratio of heavy to regular water as the Earth's oceans do. Yet, the studies of chemical composition of Halley's Comet and Comet Hyakutake conducted in the 1980s revealed that these comets have greater ratio of heavy water to regular water than the oceans on the Earth have.

Following this discovery, scientists were seriously considering removing comets from the list of our potential "water suppliers." However, the situation changed in 2000, when scientists found indirect evidence that Comet LINEAR could have the right amount of deuterium. Then, in 2011, astronomers confirmed that comet Hartley 2 contains water which is very similar to the water in the Earth's oceans.

Astronomers are not certain whether many other comets have water with the right amount of deuterium. However, the data on these two comets suggests that we should not just discard the idea of comets delivering significant amounts of water to the Earth. After all, both asteroids and comets might have made our planet "water-rich" several billion years ago.

So, the next time you turn on the tap, take a moment to reflect on the extraterrestrial origin of its flowing water. Then, you might ask yourself another interesting question: If the human body is made up mostly of water, can we say that our bodies to a certain degree are also extraterrestrial?

1.5. Can comets and meteorites make life contagious?

Some theories suggest that life on the Earth might have extraterrestrial origin because the young Earth did not have conditions suitable for life. During the Late Heavy Bombardment period, which lasted between 4.1 and 3.8 billion years ago, the Earth was heavily bombarded by comets and asteroids. We know this because we can see the Moon's surface covered with numerous ancient impact craters.

Frequent disastrous collisions between the Earth and large asteroids or comets, combined with the immense heat released during the collisions, would have made the Earth's surface uninhabitable at that time. It would be highly unlikely for any life forms or organic material to survive on the Earth under such hellish conditions. However, scientists found evidence suggesting that life may have already existed on the Earth about 3.83 billion years ago, when space debris still "mercilessly" bombarded the planet. Could that ancient life have been brought to the Earth from somewhere else?

We can interpret the idea of the extraterrestrial origin of life on the Earth in two different ways. One is that life emerged on the Earth from extraterrestrial organic material (the material that was brought to the Earth from somewhere else). Another is that life emerged somewhere else in the Solar System or beyond the Solar System and that it was then brought to the Earth.

Let us take a look at the first possibility. Astronomers have already confirmed that organic matter is relatively common in outer space. By organic matter, we mean matter composed of compounds that contain carbon (all living organisms on the Earth are carbon-based). Numerous organic compounds, including amino acids, also have been identified in some meteorites found on the Earth. (Amino acids are the building blocks of proteins, which are the primary constituents of living cells.)

For example, Sun Kwok and Yong Zhang, professors from the University of Hong Kong, discovered some interesting complex organic compounds in interstellar space. The chemical structure of these compounds is similar to those of coal and petroleum. Kwok and Zhang confirmed that an explosion of a massive star could have produced such complex organic matter in the course of a few weeks.

Interestingly, the complex organic matter produced in dying stars is similar to some complex organic matter found in meteorites on the Earth. We can explain this by the fact that stellar explosions eject that matter into the interstellar space. In the past, some of this matter could have mixed with the disk of gas and dust that later formed the Solar System. That organic matter was most likely destroyed in the harsh conditions of the young Earth, but it could have survived in comets and asteroids.

It is possible that asteroids and comets also created conditions for the emergence of life from that extraterrestrial organic matter. During the Late Heavy Bombardment, collisions between the Earth and large asteroids and comets produced huge impact craters on the Earth. These craters contained water and organic matter from the impactors (comets or asteroids), and they could have been the perfect places for the emergence of the first simple life forms from extraterrestrial organic matter.

Sankar Chatterjee, a professor of geosciences from Texas Tech University, proposed such a scenario of life emergence on the Earth. According to Chatterjee, very large meteorites created impact craters on the Earth of about 560 kilometers (350 miles) in diameter. Filled with organic matter and water, these craters became the perfect places for life's emergence. Moreover, the large meteorites could have pierced through the Earth's crust, producing volcanically active geothermal vents that could have helped to "cook" a primordial soup.

Obviously, the only way to confirm this theory would be to create life forms in lab settings that would be very similar to the environment that existed in the ancient "proto-life" impact craters. Yet, so far, no lab experiments have created life in such conditions.

Another possibility could be that life emerged somewhere else in the Solar System or beyond the Solar System and that it was then brought to the Earth. Indeed, some life forms found on the Earth can survive in extreme conditions. For example, certain kinds of bacteria can grow at temperatures as high as 113 degrees Celsius; some microbes can be well-preserved in liquid nitrogen (at a temperature of -196 Celsius) and they can survive high dosages of ionizing radiation as well as high pressure.

Bacteria make up about 30 percent of the Earth's biomass, and bacterial spores could be the prime candidates for space travel. Bacterial spores allow bacteria to exist in a dormant state under harsh conditions with no nutrients available. One study demonstrated that bacterial spores could survive in the outer space for a few years, as long as they were sheltered from solar ultraviolet radiation. Let us remind you that such conditions do exist inside comets and asteroids.

When astronomers studied the radiation emitted by the Halley's Comet dust particles, for example, they discovered some patterns that could have been produced by bacteria heated to high temperatures. However, we will have to wait for future direct studies to confirm if bacterial spores actually hide inside comets.

Are there any other ways (besides comets and asteroids) to deliver life forms to the Earth? Could aliens from other worlds bring some life forms to our planet unintentionally? After all, our modern space technology is already capable of transporting life forms from the Earth to Mars, even though life transport is not part of our space programs.

For example, when NASA specialists were preparing the *Curiosity* rover for its journey to Mars, they followed a special NASA protocol to sterilize the rover. This protocol was designed in the middle of the 20th century in order to comply with the United Nations Outer Space Treaty that prohibits the contamination of extraterrestrial worlds by Earth life, including terrestrial bacteria and other microorganisms.

The NASA specialists baked the *Curiosity* rover at 230 degrees Fahrenheit and cleaned it with alcohol. Surprisingly, their efforts were not successful. Before NASA launched the "sterilized" *Curiosity* rover into space, scientists swabbed it and checked to see if any types of organisms survived the rigorous cleaning process. They found 65 species of bacteria residing on the Curiosity and "ready to go to Mars!" Most of them were related to the genus Bacillus.

In the lab settings, scientists exposed these species to dehydration, UV radiation, pH extremes, and very low temperatures. Almost 11 percent of the 377 strains survived more than one of these harsh conditions.

It is uncertain if the microbes could survive on the surface of Mars, but it looks like the Curiosity rover could "give them a ride" all the way there. If any of these organisms survived the space trip, as well as harsh conditions on Mars, they could mutate over time. Then, scientists might later discover them on Mars without realizing their terrestrial origin.

Generally, the theory that proposes that life on the Earth originated beyond our planet and then was transported to the Earth is called the panspermia theory. Search for bacteria riding inside comets and asteroids could be the first step in any attempt to confirm the panspermia theory.

This theory does not explain how life originated anywhere else, away from the Earth. However, if it turns out to be correct, and comets and asteroids are indeed able to spread life across planetary systems, then we could ask some other interesting questions: Could comets deliver the same life material to different planetary systems?

If the answer to this question is positive, then there is a slight chance that we might have some distant relatives living on planets that orbit other stars.

Part II. The Moon

Apollo 17, the last mission to the Moon, took this image of the Moon. The right third of the picture displays part of the Moon's far side, which is never observed from the Earth. Image Credit: NASA.

2.1. Is the Moon stretching your body every day?

We are accustomed to think of gravity as an attractive force. For example, the Sun's gravitational pull holds all the planets, comets, asteroids and many other cosmic objects in the Solar System. However, gravity can also deform or even destroy cosmic objects, and gravitational interaction between the Earth and the Moon demonstrates how gravity can do that.

When we observe the Moon from the Earth, we can distinguish two sides of the Moon. The *near side* is the side of the Moon that faces the Earth, and it is the closest to the Earth. The *far side* is the side of the Moon that is the farthest from the Earth. The far side of the Moon always faces away from the Earth. If we were located on the Moon, then we could also distinguish two sides of the Earth: the *near side* of the Earth that faces the Moon and the *far side* of the Earth that faces away from the Moon.

The near side of the Earth is about 6,400 kilometers (4,000 miles) closer to the Moon than the center of the Earth is. According to Newton's Universal Law of Gravity, the Moon's gravity pulls on the near side of the Earth more strongly than on the Earth's center, producing a tidal bulge on the side of the Earth facing the Moon.

Because the far side of the Earth experiences a weaker gravitational attraction toward the Moon (as compared to the near side of the Earth), the inertia stretches the far side, shaping another tidal bulge. On the near and far side of the Earth, the planet's solid surface rises with the tidal bulges. Since water can be deformed more easily than the solid parts of the Earth, the oceans also flow into tidal bulges, producing our very noticeable tides.

This tidal force is caused by the difference between the Moon's two gravitational pulls exerted on opposite sides of the Earth. This statement is true for the Earth and for any object located on the Earth. The greater the difference in distance between the side of the object facing the Moon and the side of the object facing away from the Moon (in other words, the larger the object is), the stronger the tidal force will be stretching the object and creating tidal bulges on it.

The tidal force exerted on the Earth by the Moon produces noticeable tidal bulges on the Earth because the Earth is very large, and there is a great distance between the near and far side of the Earth. However, any tidal effects on the human body located on the Earth caused by the Moon are extremely small (practically unnoticeable) because of our small size.

2.2. Can the Moon make life on the Earth more complex?

You would probably have never thought that the Moon's influence on the tilt of the Earth's axis of rotation, which is also known as its obliquity, could make Earth's life forms more complex, but it actually might do just that.

The Earth's axis of rotation makes a 23.5 degree angle with the perpendicular to the orbital plane of our planet. This tilt of the axis of rotation is the reason that the Earth has a cycle of seasons. For example, when the Earth's northern hemisphere is tilted toward the Sun, it receives a greater amount of solar energy. As a result, we have summer in the northern hemisphere and consequently winter in the southern hemisphere.

The tilt of the Earth's axis of rotation shifts between 22.1 and 24.5 degrees over a period of approximately 41,000 years (as we mentioned earlier, the current tilt is 23.5 degrees, and it is slowly decreasing). This is called the *Milankovitch cycle*. Because of this cycle, the seasons on the Earth periodically become exaggerated to a certain degree. Greater tilt means colder winters and warmer summers; less tilt means milder winters and cooler summers.

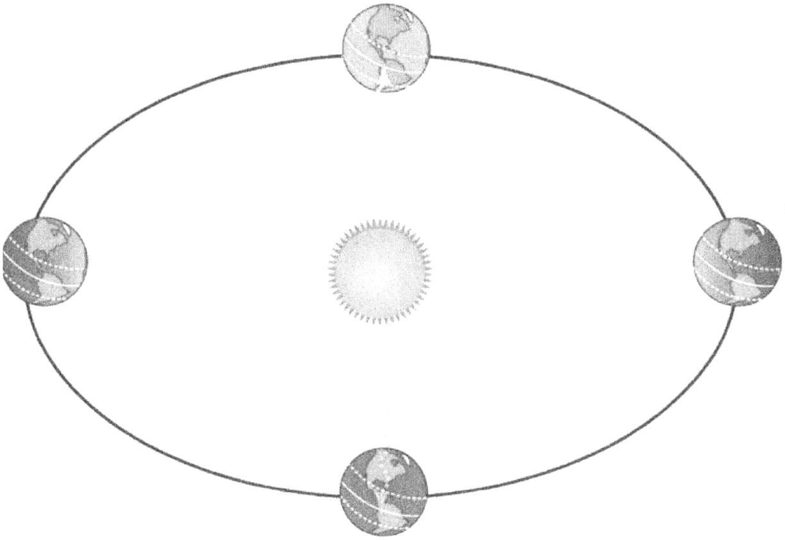

Image 2.1. Cycle of seasons on the Earth. The primary cause of the cycle of seasons on the Earth is the tilt of the Earth's axis of rotation. When the North Pole is tilted toward the Sun, the northern hemisphere experiences summer. Six months later, the South Pole tilts toward the Sun, and the southern hemisphere experiences summer. Seasons in the two hemispheres are reversed. Image Credit: NASA.

The gravitational pull of the Moon works as an anchor that prevents the tilt of the Earth's axis of rotation from wobbling too much. If the Earth did not have the Moon, the tilt of Earth's axis of rotation would be changing more drastically over the course of many thousands and millions of years.

For example, Mars does not have a large satellite like our Moon that would keep the Martian axis of rotation stable, and the tilt of the Martian axis of rotation wobbles as much as 10 degrees from its current 25-degree angle (probably every 120,000 years). As a result, the Martian climate changes so much that the ice in the polar regions of Mars periodically disappears, and then Martian ice caps move toward the Martian equator!

Changes in tilt of the Earth's axis of rotation affect our climate as well. For example, a change as small as 1.3 degrees in the tilt of the axis of the Earth's rotation could trigger an ice age on our planet. Any further increase in the Earth's axial tilt would cause dramatic changes in the Earth's climate.

Because the Moon's gravitational pull stabilizes the axis of rotation of the Earth, it also helps keep the Earth's climate more stable. And the steady climate supports conditions that favor the evolution of complex life forms. In this way, the Moon indirectly supports the complexity of life on Earth.

Could the complexity of life on the Earth be possible without the Moon? Jason Barnes and his colleagues from the University of Idaho argue that life on the Earth could do just fine without the influence of the Moon's gravity. Their studies show that without the Moon's gravitational pull, the tilt of Earth's axis of rotation would vary 10 to 20 degrees over a half a billion years. That is, the gravitational pull of other planets—particularly Jupiter—in the Solar System could have kept our planet more or less stable.

According to Barnes, a 10-degree change would have some negative effects on our climate, but it would not prevent the development of more advanced life forms on the Earth. The same should be true for other planetary systems. That is, Earth-like extrasolar planets (planets orbiting other stars) probably do not need large moons in order to have a relatively stable climate.

However, it is nice to know that our Moon adds some extra stability to our own planet, helping complex life forms to survive.

Part III. Planets

This image of the Sun and the major planets was created using a few data sets from various planetary missions. Image Credit: NASA/JPL.

3.1. Can the positions of planets in the sky determine our fate?

Modern science tells us that there is no connection between our fate and the positions of planets in the night sky. Yet, you can easily find various astrological predictions posted on the internet that tell you what is going to happen to you today, tomorrow, next week or next month. When you read any astrological predictions, please keep in mind that astrology is very different from astronomy.

Astrology is a belief system that claims that positions of celestial objects affect our fate and that one can predict the future from the positions of the Sun, stars and planets in the sky. We can trace the origins of astrology back centuries and even millennia to our ancestors' simple observations of the night sky.

When ancient astronomers systematically observed the night sky, they discovered that some patterns of celestial objects coincided with the change of seasons on the Earth. They assumed that the change of seasons was not the only phenomenon on our planet that was linked to the patterns of the night sky.

Astrology was born when ancient astronomers came to believe that positions of planets and stars in the night sky could predict major disasters and other events affecting human life. The earliest astrological studies did not predict the future for individual people. Instead, they tried to forecast such major events as wars, epidemics of dangerous diseases, earthquakes and floods. Over time, astrological studies gradually expanded to predict events in the lives of kings and royalty. The ancient Greeks then expanded astrology to predict events in the lives of ordinary citizens.

In ancient times, astronomy and astrology were intertwined because, very often, the motive for keeping observational records was astrological. Ancient astronomers were making astronomical observations and developing astrological predictions, and there was no clear divide between astronomy and astrology. Astronomical observations of the past, including ancient Chinese records of celestial events, from which the fortunes of entire dynasties were divined, are now of great value in the study of historical eclipses, novae, and comets.

Scientific discoveries made by Galileo Galilei, Sir Isaac Newton and other scientists transformed astronomy into a truly scientific discipline that relies on the scientific method and the laws of physics. The scientific method can be described in a few words: "I don't believe anything without a proof. I accept a theory or a hypothesis as the correct one only when observations and/or experiments confirm that it is correct." When scientists make any statement and propose any hypothesis or theory, they have to prove that their ideas are correct. And the proof has to come from observations and experiments.

Do you ever hear about astrologists collecting data to verify that their astrological predictions turned out to be true for a large number of people? We are certain that you did not. Astrologers use various complicated astrological diagrams and mathematics in their predictions, but they never collect the data to confirm the accuracy of those predictions. Today, astrology is not astronomy anymore, and it does not belong to the scientific realm. Astrology falls into the category of pseudoscience because it does not rely on the scientific method.

If no one ever proved that astrological predictions worked fine for millions of people, then why does astrology remain a part of many people's system of beliefs? Astrology continues to exist because it is rooted in humanity's natural desire to reduce the uncertainty of whatever the future can bring to us. We are afraid of uncertainty. The idea that someone can tell us what might happen tomorrow makes us feel more comfortable in our uncertain world.

3.2. Can Mars and the Moon throw rocks at our planet?

Sometimes Mars and the Moon throw rocks at the Earth. We know that with certainty because more than 100 Martian meteorites have been found on the Earth so far. This is not a large number though. Overall, about 1 out of 500 meteorites found on the Earth originated on Mars or the Moon.

What kind of event could send those rocks to the Earth? It turns out that these meteorites are the result of other cosmic impacts. When a large asteroid or comet smashes into Mars or the Moon, it can blast off pieces of rock from their surfaces. Ejected by the impact, some rocks can fly into interplanetary space and travel all the way to the Earth. It takes millions of years for Martian rocks to travel the distance to reach the Earth. As they fall through the Earth's atmosphere, air friction destroys some of the rocks, but others reach our planet's surface.

For example, in 2011, a Martian rock was found in the Sahara desert. It was designated Northwest Africa 7034 (NWA 7034). Scientists informally call it "Black Beauty." NWA 7034 weighs about 0.32 kilogram (11 ounces), and its chemical composition is very different from that of any other Martian meteorite found on the

Earth. NWA 7034 contains about 10 times more water than any other Martian meteorite. Moreover, it contains organic carbon! Scientists determined that NWA 7034 is about 2 billion years old. By studying this rock, scientists can get some insight into conditions on the Martian surface; for example, what Martian geological activity and atmosphere were like 2 billion years ago.

Martian and lunar meteorites found on the Earth make a wonderful present for scientists. Just think about it: instead of sending astronauts or unmanned space probes to Mars or the Moon in order to collect samples of their material, we can just find and study their rocks here, on the Earth.

3.3. Is Jupiter guarding our planet?

Imagine a giant hurricane that lasts for more than three centuries, rotates every six days, and can swallow three planets the size of the Earth. Although it sounds like science fiction, it is real. It is called the Great Red Spot and it is a storm in Jupiter's atmosphere. Driven by Jupiter's internal heat, it gives us some idea of the grand scale of events on Jupiter.

Jupiter is the largest and the most massive planet of the Solar System. Its mass is about 318 times greater than the mass of the Earth. Because Jupiter is so massive, its gravitational pull is also very strong. Jupiter's enormous gravity can change the orbits of comets that travel from the outer part of the Solar System toward the Sun. Some of these comets even end up falling into Jupiter.

In July 1994, astronomers and the public were treated to a remarkable show as comet Shoemaker-Levy 9 experienced the power of Jupiter's gravity. Generally, comets can be relatively easily broken into pieces because they are composed of ices. When comet Shoemaker-Levy 9 got too close to Jupiter, the tidal effect produced by Jupiter's gravity broke the comet into approximately 20 fragments. Then, Jupiter's gravity pulled these fragments toward the planet, making each of them crash into Jupiter and produce huge explosions in its atmosphere. The largest explosion was greater in size than the Earth!

Image 3.1. Jupiter. This true color mosaic of Jupiter was constructed from images taken by the camera onboard NASA's Cassini spacecraft in December of 2000, when the spacecraft was at its closest approach to Jupiter at a distance of about 10 million kilometers (6.2 million miles). The image clearly shows the Great Red Spot. Image Credit: NASA/JPL/Space Science Institute.

This event supports the traditional view that Jupiter acts like a guard destroying or deflecting incoming space debris (mostly comets) away from the inner Solar System, which is the home to the Earth. However, we should not put too much trust in Jupiter as a "loyal" defender. Sometimes, Jupiter does not protect our planet from dangerous comets and asteroids. Sometimes it can actually do quite the opposite.

For example, in 1770, Comet Lexell flew only a million miles from the Earth. The comet came from the outer Solar System, and it did not have any "intention" of getting so close to the Earth. However, when the comet was passing Jupiter, Jupiter's gravity pushed the comet into a new trajectory toward the Earth.

When it comes to asteroids—irregular-shaped rocky objects—we should also have little faith in Jupiter's "good behavior." Located between the orbits of Mars and Jupiter, the majority of asteroids never leave that region of the Solar System. However, Jupiter's gravity occasionally makes asteroids collide, pushing some of them toward the Earth.

It looks like Jupiter can act both ways. It can shield the Earth from comets or it can throw comets and asteroids at the Earth. However, we should not hurry into being disappointed with Jupiter. In the past, Jupiter certainly played an important role in cleaning the Solar System of most of its dangerous space debris.

The young Solar System was filled with comets and asteroids of various sizes, and collisions between this space debris and the Earth were very frequent. At that time, Jupiter and Saturn threw a lot of space debris out of the Solar System, toward its very outer edge, making our inner Solar System a much safer place.

3.4. What happens when planets refuse to go in circles?

In 1687, Sir Isaac Newton published his famous work the Philosophiae Naturalis Principia Mathematica. In his book, Newton formulated the Universal Law of Gravity and explained that the Sun's gravity is responsible for the orbital motion of planets.

According to the Universal Law of Gravity, the strength of the Sun's gravitational pull is directly proportional to its mass, and it is inversely proportional to the square of any distance from the Sun. Because the Sun is the largest and the most massive object in the Solar System, its gravitational pull is very strong. Yet, the farther away we move from the Sun, the weaker the Sun's gravitational pull is.

Ultimately, at a distance of approximately 2 light years from the Sun or 9 trillion kilometers (6 trillion miles), which is almost halfway to the nearest star called Proxima Centauri, the Sun's gravitational pull becomes so weak that we can consider it as almost non-existent.

If the Sun's gravitational force suddenly ceased to exist, the planets would travel in a straight line at a constant speed and fly away from the Solar System. It is the force of the Sun's gravity that constantly deflects the planets toward the center of the Sun. Because the planets move very fast, they do not fall toward the Sun. Instead, they orbit the Sun moving in near-circular paths.

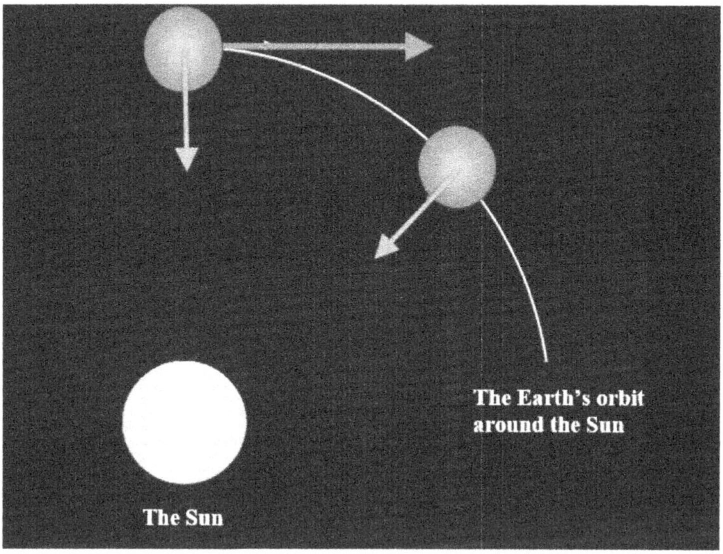

Image 3.2. The Earth would move in a straight line if no other force were acting upon it. The green arrow in the image points in the direction of such a motion. The Sun's gravity is directed toward the center of the Sun. This force deflects the path of the Earth away from its straight-line motion in to a near-circular orbit around the Sun.

Planets are also massive objects. Of course, they are not as massive as the Sun and their gravity is not as strong as the gravity of the Sun. Nevertheless, they do exert gravitational pull on other objects in the Solar

System. And because they exert a weak gravitational pull on one another, their orbits can change as a result.

Theoretical calculations and computer simulations can predict the changes in orbital motion of the planets; however, their predictions come with great uncertainties because so many planets are involved. Some astronomical studies propose that the planets will continue to move in stable orbits around the Sun at least for the next 40 million years. After that, their orbits might begin changing. As a result, one day in the distant future the Earth might collide with Mercury or Mars!

According to Jacques Laskar, a French astronomer, such a catastrophe might be triggered by Mercury. Laskar ran a large number of computer simulations developed to predict the future of the Solar System. In one or two percent of the simulations, the orbit of Mercury became very elongated because of the gravitational pull of Jupiter. Such a change in Mercury's orbit would destabilize the inner Solar System. It would increase the gravitational interactions between Mercury, Venus, Mars and the Earth, leading to a catastrophic collision between the Earth and either Mercury or Mars, or a collision between Mercury and Venus.

Gregory Laughlin, an American astronomer, and Konstantin Batygin, a Russian-American astronomer, confirmed the possibility of this scenario in the computer simulations. In the case of a collision between the Earth and Mars, the impact would heat up the Earth to thousands of degrees, and an ocean of lava would cover our planet's surface.

Based on the number of simulations, the probability of such a collision would be less than two percent. Besides, we still have many millions of years before we would have to worry about it.

3.5. Rogue planets: why are they not welcome in our Solar System?

In 2011, David Nesvorny, a researcher from the Southwest Research Institute in Boulder, Colorado, proposed that when the Solar System was very young, it should have had a fifth giant planet (in addition to Jupiter, Saturn, Uranus and Neptune). This hypothetical fifth giant planet was almost as big as Neptune. As it was orbiting the Sun, it passed too close to Jupiter, the most massive planet of our Solar System. When this happened, Jupiter's gravity gave that planet a "push" that was strong enough to send it into interstellar space.

Today, that planet might be traveling somewhere at distances of hundreds or thousands of light-years away from the Solar System. If this theory is correct, then the fifth giant planet of the Solar System has become a rogue planet. Rogue planets are the planets that do not orbit stars. Instead, they travel through the Milky Way Galaxy on their own. They are also known as free-floating planets.

Some rogue planets might have formed as lone objects. Other rogue planets most likely formed in planetary systems. Over time, gravitational interactions with other planets pushed them out into interstellar space (similarly to what might have happened to the fifth giant planet in our Solar System).

A planet can be "tossed out" of its planetary system for other reasons as well. For example, when a star grows old and expands, it transforms itself into a red giant. During this transformation, the star can push planets into interstellar space. Alternatively, when two stars pass each other closely enough, the gravity of one star can pull on some planets orbiting another star and make them leave their home system. Sometimes, dense clouds of gas and dust drifting in the interstellar space can behave as "intruders," tearing planets out of their planetary systems.

Because rogue planets drift in space by themselves and are not lighted by stars, astronomers have to use special techniques to find them. They look for the planet's infrared radiation, or they use a technique called *gravitational microlensing* that relies on gravity's ability to bend light. When a massive rogue planet passes between the Earth and a star, the planet's gravity acts like a lens. It curves the star's light, and astronomers on the Earth can observe this effect. Because the gravity of an object depends on its mass, this microlensing effect is more noticeable for very massive planets.

Some of the interstellar objects identified as rogue planets may not be real rogue planets. Brown dwarfs (objects that are bigger than giant planets, but not big and massive enough to sustain the nuclear reactions in their interior the way stars do) also wander in the darkness of interstellar space. Some rogue planets may be planets that actually do orbit stars, only they are located at great distances from their parent stars.

The two techniques that we described above allow astronomers to discover rogue planets that are at least as massive as the planet Jupiter. So far, astronomers have discovered a very small number of rogue planets. Nevertheless, they estimate that the Milky Way Galaxy could have more rogue planets than stars. In other words, there might be a few hundred billion rogue planets traveling on solitary paths in our galaxy!

If the number of rogue planets is really so big, then they might be separated by distances that do not exceed a few light-years. Considering that the distance to the nearest-from-the Sun star, Proxima Centauri, is about 4.3 light-years, some rogue planets might be located rather close to us!

So far, astronomers have not discovered any rogue planet located too close to the Solar System. This does not mean that rogue planets do not lurk somewhere close by. We simply might not be aware of them because they are not easy to detect, especially if they are small.

What makes things more interesting is that rogue planets could actually harbor life. David Stevenson, a planetary scientist from California Institute of Technology, proposes that some of these planets might have underground oceans or even some liquid water on their surfaces together with some exotic atmospheres that could support life.

The existence of livable rogue planets in the Milky Way Galaxy would open new horizons for scientists who search for life beyond the Earth. Would it not be great if one of those planets visit our Solar System so that we could study life forms inhabiting it without leaving our own planetary system?

In most scenarios, though, a visit of rogue planet would not be good news for the planets of our Solar System. If any of the nearby rogue planets were on its way toward the Solar System, then the first "thing" it would encounter approaching the Solar System would be the Oort Cloud, a spherical formation composed of billions of comets and other icy objects that orbit the Sun.

The distances between the objects of the Oort Cloud might be great enough to let the rogue planet pass through the Cloud without wreaking havoc on it and without sending a large number of comets toward the Sun. The planet might also cross the Kuiper belt, another region of space filled with icy objects. If the "visiting" rogue planet kept moving through the Solar System, its gravity could change the orbits of the planets in the Solar System, making them more elliptical or unstable.

If the Earth's orbit became more elliptical, we would experience much hotter and shorter summers along with much cooler and longer winters. This would have a negative impact on the agriculture all over our planet. The situation would only get worse should the rogue planet disturbed the orbital motion of the Earth any further.

Rogue planets are very interesting objects, but it would be best if we had an opportunity to study them while they stay away from our Solar System.

Part IV. The Sun

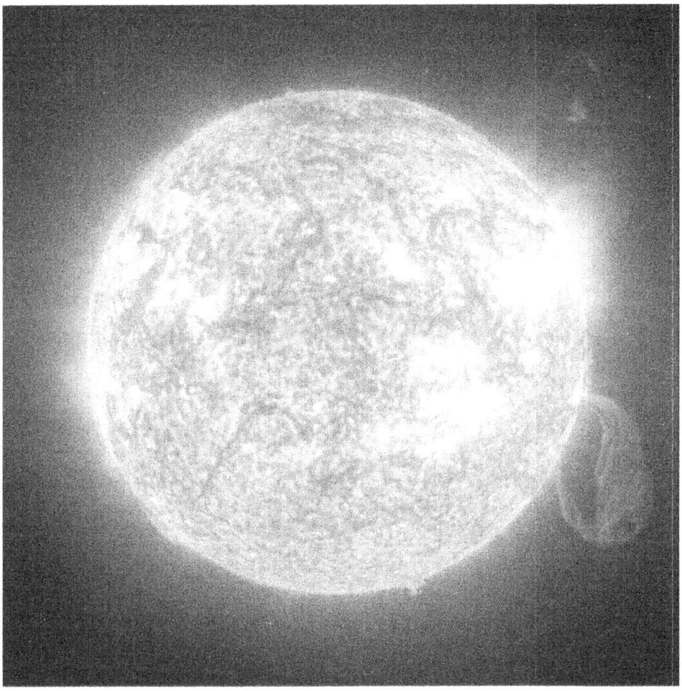

The Sun. Image Credit: SOHO - EIT Consortium, ESA, NASA.

4.1. Can we travel to the past without a time machine?

The speed of light is the greatest speed that can be achieved in our Universe, and it is truly enormous. Light travels at a speed of about 300,000 kilometers per second (186,000 miles per second). If you travelled along the Earth's equator at the speed of light, you would complete your trip in about 0.13 seconds!

A distance between any two locations on the Earth is not large enough for us to notice that light travels at a finite speed. For example, if lightning strikes at a distance of 1 mile away from you, the flash of light from the lightning will reach you within a tiny fraction of a second (a little more than one-millionth of a second). Light would travel the distance between New York City and Los Angeles in about 0.013 seconds.

However, when it comes to the Solar System and beyond, the distances are large enough to make it obvious that light moves at a finite speed. The sunlight that leaves the Sun has to travel a distance of about 150 million kilometers (93 million miles) before it reaches the Earth. Because light travels this distance in about 8 minutes, we see the Sun in the sky the way it looked like in the past, 8 minutes ago. (**Warning**: make sure that your eyes are protected with special dark filters when you look at the Sun.)

Moreover, the sunlight that reaches the Earth right now came into existence in the distant past. It was produced in the core of the Sun more than a million years ago in the form of gamma rays. Most of that radiation was transformed into visible light while it travelled from the Sun's core to the Sun's visible surface, and then to outer space.

Distances between stars are much greater than the distance between the Sun and the Earth, and it takes more time for light to travel between stars than it takes the light to travel inside the Solar System. When we look at any star in the night sky, we see the image of that star created by the light that left it a long time ago. Therefore, we see the star the way it looked in the past.

For example, Proxima Centauri, the nearest star to the Sun, is located at a distance of about 4.3 light years from the Sun. The light travels from Proxima Centauri to the Sun and the Earth in 4.3 years. When we look at Proxima Centauri, we see it the way it looked like about 4.3 years ago. Polaris, the pole star, is located further away at a distance of about 431 light years from the Solar System. On the Earth, we see Polaris the way it looked like about 431 years ago!

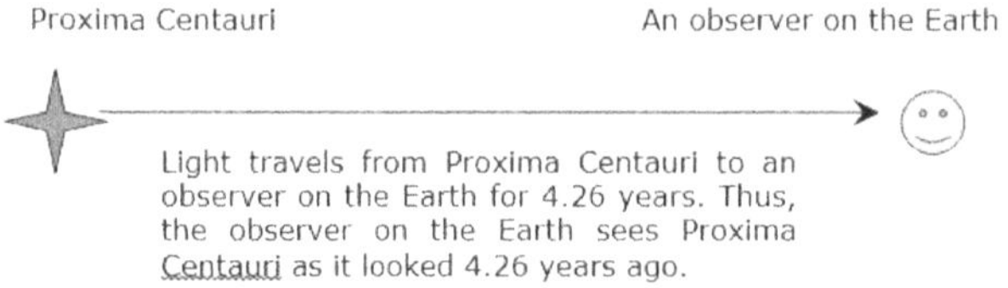

Ima

ge 4.1. Observing Proxima Centauri from the Earth.

When astronomers use powerful telescopes to observe very distant galaxies, they see those galaxies as they looked billions of years ago. This is because the light from those galaxies travels in space for billions of years before it reaches the Earth. In other words, when we look out into space, we look back in time.

4.2. Can the Sun fill our sky with clouds and send lightning to our planet?

At any moment, about 60 percent of the Earth is covered by clouds. They are composed of tiny drops of water or ice crystals that settle on dust particles in the Earth's atmosphere. Clouds form when the water vapor in the air condenses into water droplets or ice crystals. Some clouds, though, are very special. They are formed by extraterrestrial "entities"—cosmic rays—and the Sun's magnetic field may regulate the quantity of these clouds in our atmosphere. Let us take a closer look at how this happens.

The Earth's atmosphere is showered with dust particles that originate from comets and asteroids. Yet, these interplanetary dust particles are not the only "offenders" of our atmosphere. Galactic cosmic rays, which are highly energetic particles from outside the Solar System, continuously bombard the Earth's atmosphere.

In 1996, Henrik Svensmark and his colleagues from the Danish National Space Institute proposed that cosmic rays might influence cloud formation in our sky. The Danish scientists pointed out that cosmic rays could trigger the formation of aerosols, tiny particles suspended in the air. This happens when cosmic rays ionize molecules in the Earth's atmosphere. The ionized molecules interact with other molecules and create aerosols. Water vapor then condenses around the aerosols, forming cloud droplets.

The cloud formation triggered by cosmic rays can be affected by the changing magnetic field of the Sun. The Sun's magnetic field is carried out through our Solar System by the solar wind, which is a stream of highly energetic charged particles from the Sun. It tends to deflect cosmic rays away from the Earth. As the Sun's global magnetic field changes in the course of approximately 11 years, during the solar cycle, its interaction with cosmic rays also changes.

The Sun's magnetic field becomes stronger during the solar maximum and wanes during the solar minimum. During the solar maximum, the Sun's magnetic field and the solar wind may act as a shield protecting the Earth's atmosphere from cosmic rays. In this way, the Sun hinders the cloud formation that might be otherwise triggered by incoming cosmic rays. During the solar minimum, greater amounts of cosmic rays reach the Earth, potentially resulting in more intensive cloud formation in the Earth's atmosphere.

But this is not all. In addition to changing the cloud cover over our planet, the Sun may also affect its atmospheric electricity. Recently, Christopher Scott and his colleagues from University of Reading in the UK, determined that the number of lightning bolts in our atmosphere increases when strong gusts of the solar wind enter the Earth's atmosphere.

The solar wind is made of slow and fast streams that propagate behind each other as the Sun rotates. Scientists discovered that in the course of 40 days that followed the arrival of a high-speed stream of the solar wind, the average number of lightning strikes in Northern Europe amounted to 422. This was greater than the average 321 lightning strikes that occurred during the 40 days following the arrival of the low-speed stream of the solar wind.

Scientists propose that the solar wind particles may penetrate storm clouds, making it easier for the clouds to discharge atmospheric electricity in the form of lightning bolts. This process can work all around the planet, not just in Northern Europe.

The Sun, the Sun's radiation and magnetic field, as well as the solar wind affect the Earth in many different ways. Because they change conditions within the Solar System and influence our planet's magnetic field and atmosphere, scientists describe their dynamics and their impact on the Earth in terms of space weather. Many government agencies, including NASA, monitor space weather.

4.3. Should we worry when the Sun turns the Northern and Southern lights on?

The aurorae, which are also known as Northern and Southern Lights, are frequently observed at high northern or southern latitudes. An aurora usually looks like luminous bands or streamers extending in the sky to altitudes of 322 kilometers (200 miles). Although we think of the aurora as a luminous phenomenon in the Earth's atmosphere, it would not exist without the Sun. The formation of the aurora requires two major players: the solar wind and the Earth's magnetosphere.

The Earth's magnetosphere is the area of space surrounding our planet that is controlled by the planet's magnetic field. It shields the atmosphere of our planet from the solar wind. The solar wind carries charged particles and a certain amount of the Sun's magnetic field. It blows at a speed of about 400 kilometers per second (250 miles per second) when it gets close to the Earth.

When the solar wind reaches the magnetosphere of the Earth, its electrically charged particles become trapped in the Earth's magnetic field. As they spiral back and forth along the magnetic field of the Earth, they enter the Earth's atmosphere near the north and south magnetic poles of the planet where they collide with oxygen, nitrogen and other atmospheric gases.

In the collisions between the solar wind particles and the molecules of atmospheric gases, the solar wind particles transfer some of their energy to the molecules of oxygen and nitrogen. The molecules of oxygen and nitrogen absorb the energy and then re-emit part of it in the form of visible light. We observe that visible light as the aurora.

Today, we can enjoy the spectacular show of Northern and Southern lights without much worry. However, the solar wind that blows at us from the Sun can become a major concern in the distant or not-so-distant future. Although we do not expect any significant changes in the solar wind, scientists are certain that the Earth's magnetic field is preparing for some major changes.

The Earth's magnetic field looks like the magnetic field of a bar magnet that has north and south magnetic poles. The intensity of the Earth's magnetic field can remain almost the same for thousands or millions of years, but it does periodically weaken and reverses its direction, flipping its magnetic poles.

Keep in mind that this process does not happen overnight. Usually the magnetic field reverses its magnetic poles in the course of a few thousand years. Sometimes, the reversal happens much faster. For example, the last magnetic reversal happened about 786,000 years ago and it lasted for less than 100 years.

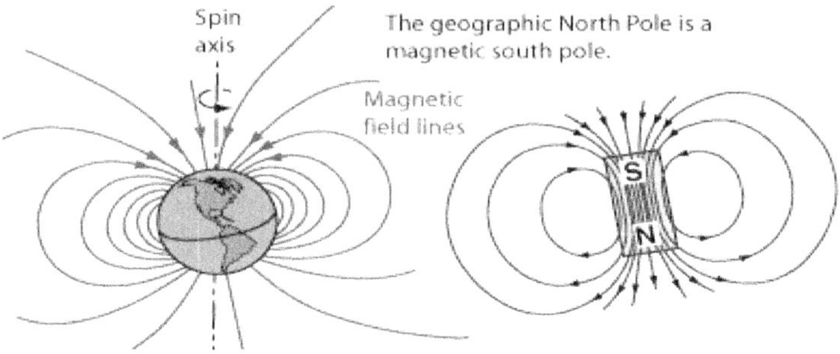

Image 4.2. Magnetic field of the Earth. Image Credit: Geological Survey of Canada.

Why would it matter? Well, it matters because currently the Earth's magnetic field is weakening. Its strength has decreased by about 15 percent over the last two centuries. This means that, most likely, the Earth's magnetic field is "heading" for another magnetic reversal. Scientists estimate that the next reversal may happen within the next 2,000 years.

From the study of previous magnetic reversals, we know that during a reversal, the Earth's magnetic field becomes very weak or might even temporarily disappear. And when this happens, the solar wind will strike the Earth's atmosphere directly, eroding our planet's upper atmosphere and "swiping" some very small fraction of our atmosphere into space.

The solar wind particles will be directly colliding with the atmosphere, making aurorae visible every night all around the planet. These energetic particles will also endanger the artificial satellites and astronauts even more than they usually do.

Furthermore, when the Earth's magnetic field weakens or disappears, severe solar storms might damage power grids on our planet, leaving people without electrical power supply for a few weeks or even a few months. In the absence of the Earth's magnetic field, solar wind particles could affect our weather and create holes in the ozone layer.

Without the Earth's magnetic field, there would be an increase in the amount of cosmic radiation at ground level. Energetic particles and radiation traveling through human body could cause various health problems, including low blood count, cataracts, radiation sickness and cancer.

Because both the Earth's atmosphere and magnetic field protect us from much of cosmic radiation (as long as we remain on the surface of our planet), the planet's atmosphere would still provide us with a significant protection even in the absence of its magnetic field. However, we would witness frequent genetic mutations and a rise in cancer rates.

So far, scientists have not found any connection between mass extinctions and magnetic reversals. And it is unlikely that the next reversal would cause any mass extinction. However, some species of birds, whales and fish do need the Earth's magnetic field for "navigation" because they sense the direction of the Earth's magnetic field, using it for migration. They would have to adapt or become extinct when the magnetic field changes or disappears.

We will be fortunate if the next magnetic reversal happens quickly. Otherwise, if it extends over centuries, then many generations of humans and all the other life forms on the Earth will have to deal with a weak and unstable magnetic field.

4.4. How does the Sun turn off GPS systems and cause blackouts on our planet?

On the morning of September 1, 1859, the British astronomer Richard Carrington observed the Sun using his telescope. He noticed a very bright light above a large group of sunspots on the Sun. On the same day, strange events occurred all over our planet: the telegraph network in the US and Europe went down; magnetized needles of compasses spun randomly, unable to point toward the Earth's magnetic poles; the aurorae were visible at high latitudes, mid-latitudes and even at near-tropical latitudes over Jamaica, Hawaii, the Bahamas, and Cuba.

In 1989, a similar event shut down power grids from the Hydro-Québec generating station in Canada. About 6 million people were left in darkness for about 9 hours. In December of 2005, another similar event interrupted communication with satellites and Global Positioning Systems (GPSs) for approximately 10 minutes. It endangered commercial airplanes that needed these systems for safe navigation.

Over time, scientists realized that these electronic disasters on the Earth were caused by storms on the Sun. Observations of the Sun's activity for the last 2,000 years revealed that solar storms come and go in cycles, and their strength and frequency change over time. Astronomical studies also found that solar storms are magnetic in nature. They are caused by the interactions between the solar matter and the Sun's magnetic field.

The Sun sends into space a fairly stable amount of energy. The average variation in the Sun's visible light is approximately 0.1 percent over the 11-year solar cycle. These variations are so minor that scientists were able to discover them only using satellites located above the Earth's atmosphere. However, the 11-year sunspot solar cycle and the 22-year solar magnetic cycle tell us that the Sun is a magnetic variable star and its magnetic field noticeably varies with time.

Solar magnetic storms are frequently accompanied by solar flares, giant explosions in the low atmosphere of the Sun, appearing near sunspots. Solar flares vary in size and strength. The biggest solar flares are called X-class flares. (**Warning:** astronomers use special astronomical instruments to observe solar flares; do **not** try to observe solar flares with a naked eye.)

A solar flare is observed as an abrupt, fast, and intense variation in brightness. It occurs when built up magnetic energy in the Sun is rapidly released in its atmosphere. The bright light that Carrington observed above a large group of sunspots on the Sun in 1859 was, in fact, the first solar flare described in astronomical literature. (Carrington was not the only astronomer who observed it. Richard Hodgson, another English astronomer, independently observed the same bright flare of light.)

Image 4.3. The Sun. A gigantic magnetic loop of hot gases forms a glowing handle on the Sun. It was caused by instabilities in the Sun's magnetic field. Image Credit: NASA.

Solar flares emit electromagnetic radiation across the electromagnetic spectrum, from radio waves to visible light to x-rays and gamma rays. They also send energetic charged particles into space. One solar flare can release the amount of energy, which is the equivalent of the energy released by millions of 100-megaton hydrogen bombs!

The intense heat of a solar flare cannot reach the Earth, and even the biggest solar flare cannot destroy the Earth. However, electromagnetic radiation and highly energetic particles from a solar flare can travel all the way to our planet. Once they reach the Earth, they can temporarily change the upper layers of the planet's atmosphere and cause disruptions in signal transmissions from a GPS satellite to the Earth.

Solar flares and coronal mass ejections (CMEs) frequently happen together because they both depend on the Sun's magnetic field. However, each can also occur in the absence of the other. Solar flares form more rapidly and they emit intense x-rays and gamma rays that travel at the speed of light and reach the Earth in approximately eight minutes. CMEs, large bubbles of magnetized gas, form slowly and they do not produce intense visible light. But they can be even larger than the Sun! It takes about several hours for CMEs to erupt into space and about 2 to 4 days to reach the Earth.

When X-class flares and CMEs are directed toward our planet, they can cause radiation storms damaging communications systems, satellites and power grids on the Earth. When a CME reaches the Earth, it sends bursts of particles and fluctuating electromagnetic radiation into the atmosphere of our planet. The effect of the electromagnetic fluctuations can propagate to the ground, blowing out transformers in power grids. In addition, a CME's energetic particles can damage electronics on board a satellite in orbit.

Although we cannot prevent solar storms, scientists at NASA and NOAA can use space technology to constantly monitor the Sun and warn us about any CME that heads toward the Earth.

4.5. How does the Sun flood the city of Venice?

For centuries, scientists have wondered whether solar cycles and sunspots could affect the Earth's weather and climate. Various scientific theories have been proposed to explain the connections between cyclic changes in the Sun and our planet's weather and climate. Some theories relate the changes in the Sun's ultraviolet radiation to the changes in the Earth's atmosphere.

For example, any variations in solar ultraviolet radiation during a solar cycle affect atmospheric ozone. An increase in solar energy reaching the Earth during the solar maximum is absorbed by ozone in the stratosphere, one of the layers of the Earth's atmosphere. This process warms up the stratosphere and changes stratospheric winds. The increase in solar radiation during a solar maximum also causes warming up of the subtropical Pacific Ocean. Combined, these changes might change weather patterns similar to that seen in a La Niña event.

In addition, solar cycles might indirectly trigger high-surge events in the North Adriatic, and these events have been linked to the flooding of Venice, Italy. Venice is usually hit by a very high tide called the *acqua alta* a few times per year. David Barriopedro and his colleagues from the University of Lisbon, Portugal, studied the hourly sea level records collected for Venice between 1948 and 2008 and compared them with the records of solar cycles. The scientists noticed that such extreme tides follow the solar cycle and that they peak during the solar maximum. Most likely, changes in the Earth atmosphere caused by solar activity were involved in this process.

4.6. Can a solar grand minimum cause another ice age on our planet?

Another important question to consider is whether the change in the total solar electromagnetic radiation during a solar cycle can cause global warming or global cooling on the Earth. In order to answer this question definitively, we would need to see records of solar radiation that extend many centuries into the past. However, the only direct measurements available are those from satellites located above the Earth's atmosphere. So, records exist since only 1978.

The Earth's climate models tell us that changes in total solar radiation during one 11-year solar cycle do not last long enough to cause global climate change on the Earth. The change in the Earth's global mean temperature during one solar cycle is estimated to be less than 0.1 degrees Celsius.

Climate models do indicate, however, that if a change of 0.1 percent in the total solar radiation persisted for a few decades, then it could change the Earth's global mean temperature by about 0.2 degrees Celsius. Accordingly, some scientists continue studying the possible effects of a few unusually long solar minima on the Earth's climate—including the Maunder Minimum and the Dalton Minimum.

The Maunder Minimum lasted from 1645 to 1715. The Little Ice Age on the Earth, a period of abnormally cold weather in Europe and North America, began around 1500, before the Maunder Minimum, and lasted until 1850.

Is it possible that the low solar activity and the decrease in the total solar radiation during the Maunder period actually triggered the Little Ice Age? Some scientists suggest that it is not simply the low solar activity during the Maunder period, but its duration that contributed to the onset of the Little Ice Age.

The Maunder Minimum is not the only unusually long solar minimum. The Dalton Minimum was a period of low solar activity that lasted from approximately 1790 to 1830. Both the Maunder Minimum and the Dalton Minimum matched periods of lower-than-average global temperatures on the Earth.

Other long periods of solar maximum or minimum occurred before astronomers began systematic telescopic observations of the Sun. To study them, scientists have used ancient records documenting naked-eye sunspot observations as well as other indirect ways to backtrack sunspot activity over the last few thousand years (sunspot activity is directly related to the Sun's magnetic activity).

The most useful observations of sunspots were documented by Chinese astronomers more than 2,000 years ago. From their records, astronomers have indirectly reconstructed suspected sunspot activity for thousands of years back. Past solar cycles also can be tracked by measuring the abundance of isotope carbon 14 that is produced when galactic cosmic rays (high-energy particles) enter the Earth's atmosphere. During the solar maximum, the solar magnetic field and the solar wind shield the Earth's atmosphere from cosmic rays, inhibiting the production of carbon 14 in the Earth's atmosphere.

The amount of solar energy that reaches our planet is only one of many factors that affect the Earth's climate. So far scientists have not reached a conclusive agreement on whether abnormally long solar maxima or minima contribute greatly to the periods of global warming and cooling that have occurred throughout Earth's history.

4.7. What does the Faint Young Sun Paradox tell us about Earth climate theories?

Interestingly, the climate theories that promise ice age conditions on the Earth due to the cooler Sun are unable to solve one famous climate puzzle. Approximately 4.5 billion to 2.5 billion years ago, the Sun was about 30 percent cooler and dimmer than it is right now. With less solar radiation reaching the Earth, our planet should have been completely frozen early in its history.

Nevertheless, geologists and paleontologists found sufficient evidence confirming that liquid surface water and even life existed on the Earth 3.8 to 2.5 billion years ago. This means that some other effects must have been compensating for the faint and cool young Sun. Scientists call this puzzle the *"Faint Young Sun Paradox."*

Possible solutions to the paradox suggest that higher concentrations of carbon dioxide, methane or ammonia in the young Earth's atmosphere may have kept it warmer. Recently, David Minton, a researcher from Purdue University, proposed a drastically different solution. His calculations show that the young Earth could have avoided the deep freeze if it was 6 or

7 percent (a few million kilometers or miles) closer to the Sun than it is now.

After all, with all the different theories proposed, the Faint Young Sun Paradox still needs an unambiguous solution.

Part V. Stars

A false-color image of globular star cluster NGC 362, a densely packed group of old stars located about 30,000 light-years away from the Solar System. In the image, which was obtained by NASA's Galaxy Evolution Explorer, we can see yellow-tinted stars that surround a white-yellow spot. The white spot is the cluster's core. It is made up of stars located so closely to one another that they cannot be seen individually in the image. Image Credit: NASA/JPL-Caltech/University of Virginia

5.1. Can a supernova explosion destroy our Solar System?

When a star similar to our Sun dies, it expels its outer layers into space. Some other stars end their lives in much more powerful explosions called supernovae. A single supernova can outshine an entire galaxy of billions of regular stars! It can remain very bright in the sky for weeks.

Image 5.1. Supernova 1994D was a Type Ia supernova discovered near the outskirts of galaxy NGC 4526. Image Credit: NASA/ESA/Hubble Space Telescope/High-Z Supernova Search Team.

Although astronomers often observe supernovae in other galaxies, supernovae are difficult to detect in our own Milky Way Galaxy because dust in the disk of our galaxy blocks our view. Scientists estimate that about one or two supernovae go off in our galaxy every century.

The last observed supernova in the Milky Way Galaxy was discovered by Johannes Kepler, a German mathematician and astronomer, in 1604. It was located at a distance of approximately 20 thousand light years away from the Earth. Using NASA's Chandra telescope, astronomers recently discovered the remains of another supernova that exploded in the Milky Way Galaxy more than a century ago.

Astronomers divide supernovae into two categories, Type I and Type II. These two categories differ by the spectrum of electromagnetic radiation emitted by the matter that they eject into space, as well as by the changes in their brightness.

A type Ia supernova is a white dwarf that explodes in a binary system. Because white dwarfs have a fairly standard mass, astronomers can predict the brightness of a Type Ia supernova. They can further compare it with the observed brightness of the supernova and determine how far away it is located. This is why astronomers also call type Ia supernovae "standard candles." Their brightness changes over time. Initially,

Type Ia supernovae demonstrate a very rapid increase in luminosity, reaching their maximum brightness over the course of about two weeks. Then, their brightness declines rapidly for about three or four weeks.

White dwarfs are extremely dense stellar objects primarily composed of carbon and oxygen. They are about Earth-size, and their mass approximately equals the Sun's mass. White dwarfs form from the collapsing stars that were formerly similar to the Sun. White dwarfs can exist for billions of years, supported by electron degeneracy pressure. Because electron degeneracy pressure does not allow electrons to become "squeezed" too close to one another inside a white dwarf, it offsets the white dwarf's self-gravity, preventing any further collapse.

However, if the white dwarf gains more material, and its mass exceeds approximately 1.4 solar masses (the Chandrasekhar limit), then its self-gravity overpowers electron degeneracy pressure. The white dwarf's mass becomes high enough so that it can begin fusing carbon into heavier elements. This process triggers a runaway thermonuclear reaction. Thermonuclear fusion happens extremely fast, abruptly releasing a tremendous amount of energy that destroys the white dwarf. This process would be similar to a hypothetical case of an explosion of a hydrogen bomb, if the bomb were approximately the size of the Earth and as "heavy" as the Sun!

In order to become a Type Ia supernova, the white dwarf has to acquire enough material from somewhere else. It needs a "donor." Most likely, it can get that extra material from its stellar neighbor. In order to have a very close neighbor, the white dwarf has to belong to a binary system where two stars orbit one another. In such a binary system, the white dwarf's strong gravity can draw material off its companion until the white dwarf becomes massive enough to ignite the runaway thermonuclear reaction.

A red giant can play the role of a companion star that "feeds" the white dwarf and makes it heavier over time. However, sometimes it takes two white dwarfs to make one Type Ia supernova. For example, in January 2014, astronomers observed as a new supernova had lighted up in the nearby galaxy M82. They named the newly discovered event SN2014J. Located approximately 11.5 million light years away from us, it was the tenth supernova that astronomers detected in 2014.

Astronomers had to do some "forensic investigation" in order to figure out what caused SN2014J. If the supernova was the explosion of a white dwarf that collected too much matter from its companion star (for example, a red giant), then the white dwarf would have been surrounded by a cloud of gas before it exploded. (This is because the transfer of stellar material from the red giant to the white dwarf is not completely "clean" process, and some of that stellar material would end up in space.) The blast wave from the supernova would have smashed into that material, emitting detectable X-rays. Because astronomers did not detect such X-rays, they concluded that space surrounding the white dwarf was clean.

Image 5.2. Supernova 2014J in Galaxy M82. Image Credit: NASA and ESA.

Then, astronomers came up with another explanation for SN2014J: if the supernova was not the destructive explosion of a white dwarf that was feeding on a red giant, then SN2014J could be the result of the collision of two white dwarfs! Astronomers theorized that as the two white dwarfs orbited one another in a binary system, they would get closer together while moving at the speed of about one million miles per hour. Their orbits would get smaller and finally the two dwarfs would merge, triggering a supernova explosion.

Type II supernovae are different from Type Ia supernova. Type II supernovae are explosions of old "supergiant" stars. When a dying supergiant star runs out of its nuclear fuel, the star's collapsing core triggers an explosion that we observe as a supernova. The brightness of Type II supernovae peaks after a week or two, and it then remains almost constant for about two months. After that, the brightness decreases abruptly.

Supernovae throw matter out into space at a speed of 15,000 to 40,000 kilometers per second (9,000 to 25,000 miles per second). While a supernova sends most of a star's material in the interstellar space, the central core of a star undergoes further gravitational contraction and might become a white dwarf, neutron star, quark star, or black hole.

A type Ia supernova can release energy that is the equivalent of 20 billion, billion, billion megatons of TNT. If a star went supernova at a distance of 1 light year away from the Earth, its shockwaves would destroy our Solar System completely!

Any supernova explosion within 30 light years of the Earth would be devastating for life on the Earth.

Scientists estimate that X-rays and gamma rays emitted by such a supernova would destroy nearly all the ozone layer in our atmosphere. Without sufficient protection provided by the ozone layer, the Sun's ultraviolet radiation would destroy much of the life on the Earth's surface and in the top layers of the Earth's oceans, seas and lakes. In addition, the radiation from a supernova would also ionize oxygen and nitrogen in our atmosphere, producing large amounts of smog-like nitrous oxide.

According to Mark Reid, an astronomer from the Harvard-Smithsonian Center for Astrophysics, a supernova within 30 light years of the Sun should occur about once every 100 million years. Today, there are no stars located that close to us that would be candidates for a supernova. One of the nearest stars that could go supernova is Betelgeuse, but it is located 640 light years away, so its explosion would not hurt life on the Earth.

Evidence for past supernovae that went off close to our Solar System might have been found on the Earth. In 1999, German researchers discovered traces of iron-60 in south Pacific sediments. Supernovae typically expel this isotope, and its presence on the Earth hints that one of our neighboring stars went supernova just a few million years ago. It was located somewhere within approximately 100 light years of our Solar System.

We should keep in mind, though, that even distant supernovae can reach out to us. Supernova shock waves produce cosmic rays, which are mostly high-energy charged particles called protons. These cosmic rays may bounce along the magnetic field of our galaxy for millions of years before they reach the Earth. Many of them find their way to our planet all the time: each square centimeter on the top of the Earth's atmosphere is bombarded by a few cosmic rays every second!

These cosmic rays do not reach the ground. They collide with atoms in the Earth's upper atmosphere, producing a shower of less-energetic "secondary" particles. The majority of secondary particles reach sea level as highly penetrating particles called muons. Approximately 10,000 muons travel through our bodies every minute, and some of them can cause genetic mutations in human body.

Because humans have been exposed to the influence of cosmic rays for millennia, we should not consider "cosmic-ray" genetic mutations as a reason for panic. Some of these mutations can be harmful, but some genetic mutations probably contribute to evolution of life on the Earth.

5.2. Exploding stars: are they good or bad for us?

If a supernova went off too close to the Solar System, it would destroy life on the Earth. However, more distant supernovae actually might have played a positive role in the formation of the Solar System and the emergence of life on our planet. Their contribution to life on the Earth is related to the changes that stars undergo in their lifetimes.

Stars are born, live their lives and then die. Their lifespan depends on their mass. Many "medium-mass" stars can exist for billions of years. Considering that humans first began to evolve hundreds of thousands of years ago, we have not been around long enough to observe the evolution of any star from its birth until its death. However, astronomers study a large number of stars of different ages and use the laws of physics in order to develop theories explaining how stars change over time.

We know from observations that stars are born in dense regions of clouds of gas and dust, which collapse into stars under the influence of gravity. A star is being born when the temperature in its central region reaches 10 million Kelvin (18 million degrees Fahrenheit). At such a high temperature, nuclear fusion reactions begin in a star's core. They convert hydrogen into helium and release energy. The outflow of that energy from the central regions of a star provides the pressure that prevents the collapse of the star under its own weight.

Image 5.3. Gas pillars in the Eagle Nebula that is located 6,500 light years away from the Earth. The gas pillars in the Eagle Nebula are columns of interstellar gas and dust. They are nurseries for new stars. The Hubble Space Telescope took this picture in April 1995. Image Credit: NASA, ESA, STScI, J. Hester and P. Scowen (Arizona State University).

The nuclear fusion reactions inside stars release tremendous amounts of energy. For example, there are more than 400 commercial nuclear reactors on the Earth. It would take about 50 million years for these nuclear reactors to produce the same amount of energy that the Sun produces in one second!

In the course of a star's life, the nuclear reactions in its core transform lighter elements into heavier elements: hydrogen is converted into helium; helium is converted into heavier elements, including carbon, etc. When stars approach the end of their lives, most of them expel stellar matter into space. Very massive stars end their lives with extremely violent explosions called Type II supernovae.

Astronomers estimate that the Universe was born approximately 13.7 billion years ago, and only the lightest elements—hydrogen, helium, and a tiny amount of lithium—were created in the Universe at that time. All the heavier chemical elements that exist in our Universe were produced inside stars and expelled into space by stellar explosions. (Elements that are heavier than iron are made in supernova explosions.) This is why we can say that our planet and our bodies are made of stellar matter.

But how did matter expelled into space by supernovae get into our Solar System and our bodies?

Well, it had to travel great distances and experience transformations before it became part of us.

All the chemical elements sent into space by supernovae mix with interstellar clouds of gas and dust, and some of these clouds collapse and give birth to the next generations of stars and planets orbiting these stars. About 4.6 billion years ago, one of these clouds of gas and dust—enriched by the chemical elements from the old exploding stars—was compressed by some kind of event. (It could have been compressed by the shock wave from a supernova.)

The Sun and the Solar System formed from a fragment of that contracting cloud. In the process, the planets of the Solar System—including the Earth—incorporated chemical elements that came from other stars of the Milky Way Galaxy.

You might be accustomed to counting your biological age from the date of your birth. However, if we had to determine your physical age as the age of the chemical elements in your body, we would have to recall that the chemical elements were created in stars and supernovae billions of years ago. This means that you (and any other living creature on the Earth) are a few billion years old!

We can say that distant supernovae can be "beneficial" because they enrich interstellar clouds of gas and dust with various chemical elements, and they also produce shock waves that compress clouds of gas and dust, "encouraging" them to begin the formation of stars.

However, we should remember that not all stellar explosions are good for us, and we have one "stellar bomb" ticking very close to us. We are talking about our own Sun. Scientists estimate that the Sun's core should contain enough hydrogen to generate energy at present rates for approximately another 4.5 billion years. After that time has passed, when the Sun has exhausted its hydrogen fuel, it will go through catastrophic changes (but short of a supernova because of the Sun's mass) that will destroy the Earth and other planets of the Solar System.

This is a bad news. The good news is that we have plenty of time to prepare for doomsday.

Part VI. Our Galaxy

This image of the Milky Way Galaxy was taken by the satellite COBE. We can see the disk and central region of our galaxy in the image. Image Credit: The COBE Project, DIRBE, NASA.

6.1. Can our galaxy cause havoc on the Earth?

Our galaxy can affect the Earth in various ways. For example, it is the home to potential supernovae that might be very dangerous when they go off too close to our Solar System. The Milky Way Galaxy also hosts hypernovae that are about ten times more powerful than typical supernovae. Hypernovae emit gamma ray bursts (GRBs) that can travel about 6,500 light-years! One of these hypernovae could have caused mass extinctions on the Earth within the last billion years.

If a GRB occurred at a distance of 1,000 light-years from the Solar System, and it was pointed directly at the Earth, then for a few seconds it would look much brighter than the Sun in our sky. The side of the Earth facing the GRB would be blasted, and life forms on that side of the Earth would be destroyed. At the same time, extremely strong winds would be blowing around the planet. Some life forms would be able to survive in the oceans and seas on the side of the Earth facing away from the GRB.

A collision of neutron stars can also produce extremely short gamma ray bursts (so far, astronomers have discovered more than 2,000 neutron stars in our galaxy). The Japanese researcher Fusa Miyake found evidence that such a gamma ray burst might have reached our planet about 12 centuries ago. Miyake discovered high levels of carbon-14 in the composition of ancient trees in Japan. The rings of the trees allowed the scientist to date the carbon-14 to the year 775 CE. High levels of beryllium-10 dated to the same period were discovered in Antarctica.

Carbon-14 and beryllium-10 are produced when a burst of radiation strikes the Earth's atmosphere. The amount of carbon-14 and beryllium-10 that formed about 1,200 years ago in the trees in Japan, as well as in Antarctica, is so high that Valeri Hambaryan and Ralph Neuhäuser, researchers from University of Jena, Germany, propose that most likely a short GRB could have sent such intense blast of radiation to the Earth.

As the Solar System moves along a near-circular path around the galactic center, it completes one revolution in about 230 million years, and it remains at the same distance of about 28,000 light years from the galactic center. Some regions of the Milky Way Galaxy have more "things happening" than other regions, and the Solar System's ride through the Milky Way Galaxy can become "bumpy" when the Solar System enters an "unsafe" neighborhood filled with interstellar gas and dust.

Presently, the Solar System travels through a local bubble, a region of the Milky Way Galaxy that has little gas (about one hydrogen atom per five cubic centimeters of space). However, in the past, the Solar System must have travelled through much denser clouds of gas and dust that measure more than 100 light years across. Such passages through clouds of gas and dust happen every few hundred million years, and they can be harmful for our planetary system.

Let's take a closer look at how they can be bad for the Solar System. The solar wind usually forms a huge magnetic shield, the heliosphere that protects the Solar System from interstellar media (gas and dust) and radiation. When the Solar System enters more dense areas of the interstellar gas, the solar wind cannot push far enough. As a result, the heliosphere can shrink all the way to the Earth's orbit.

According to Alex Pavlov, a researcher from the University of Colorado in Boulder, this can allow the dust from outer space to block sunlight and trigger a global ice age on the Earth. In addition, the buildup of hydrogen from outer space in the Earth's upper atmosphere would change atmospheric chemistry and create a cloud layer that would reflect sunlight and further contribute to the ice age cold, putting the Earth into a deep freeze, with the planet completely covered in snow and ice.

Interstellar clouds of gas and dust could also send some exotic microorganisms and more organic material toward the Earth and the other planets of the Solar System. Moreover, the gravitational interaction between the Solar System and the interstellar clouds of gas and dust can push more comets and asteroids on a path of collision with the Earth.

The Solar System's trip through one of the galaxy's spiral arms can be a dangerous endeavor. Miroslav Filipovic, Evan Crawford and Nick Tothill, researchers from University of Western Sydney in Australia, found a correlation between the six mass extinction events that happened on Earth in the past and the times of the Sun's historical passages through the spiral arms of our Galaxy. The extinction of the dinosaurs about 65 million years ago was one of those events.

Their finding indicates that the galaxy's spiral arms might cause catastrophic events on the Earth in different ways. Our galaxy's spiral arms harbor denser than normal clouds of gas and dust where many new stars are being born. Some stars also end their lives in catastrophic explosions. Comets, asteroids and explosions of nearby stars can trigger dangerous events. Some of them can also increase volcanic activity on the Earth.

But that's not all. While our Solar System travels inside the Milky Way Galaxy, the galaxy itself travels in intergalactic space at a speed of more than 150 kilometers per second (93 miles per second)! During this travel, highly energetic intergalactic cosmic rays bombard our galaxy.

As long as the Solar System remains in the disk of the Milky Way Galaxy, the magnetic field of the galactic disk protects the Solar System from most of these intergalactic cosmic rays. However, as the Solar System moves inside the Galaxy, it dips 200 light years above and 200 light years below the galactic disk every 60 million years. And when the Solar System wanders away from the galactic disk, it becomes more exposed to harmful intergalactic cosmic rays.

Today, the Solar System is already located north of the galactic disk, and it is currently still moving farther away from the area where the galactic magnetic field can protect us from intergalactic cosmic rays. In a few million years in the future, the intergalactic cosmic rays will likely become a real problem for future generations of humankind.

Part VII. The Universe

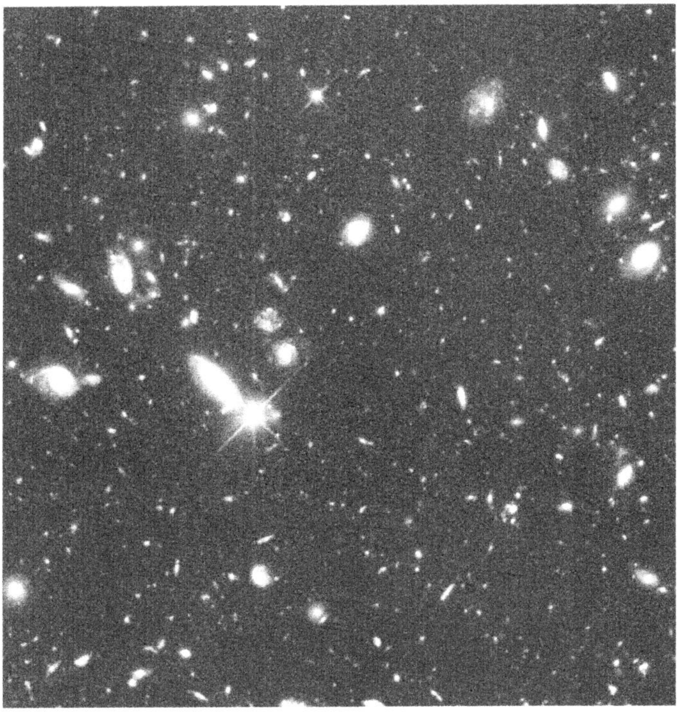

Hubble's view of the Universe reveals distant galaxies that have never been seen before. This "deepest-ever" panoramic image of the Universe is called the Hubble Deep Field (HDF). It was made with NASA's Hubble Space Telescope. Credit: Robert Williams and the Hubble Deep Field Team (STScI) and NASA.

7.1. What can dark matter and dark energy do to us?

Imagine if you had to walk into a dark room filled with furniture. You would most likely feel uncomfortable walking in the dark and bumping against the things you could not see. Yet, scientists suspect that we travel "in the dark" in this way all the time. Recent astronomical discoveries tell us that the Earth, the Solar System and the Milky Way Galaxy might be on a journey through invisible "stuff" scattered throughout the Universe.

The visible part of our Universe consists of billions of galaxies and all the other visible objects that populate galaxies, including planets, comets and asteroids, stars, brown dwarfs and clouds of interstellar gas and dust. All these cosmic objects emit various types of electromagnetic radiation (visible light, infrared radiation, ultraviolet radiation, etc.), and they are made of ordinary matter—protons, neutrons and electrons held together in atoms. This ordinary matter makes up less than five percent of the mass of our Universe. The rest of it is dark matter and dark energy.

Interestingly, if you read any physics textbook, you will find that more than 90 percent of the textbook is devoted to the discussion of the physical laws that rule ordinary matter, and less than 5 percent of the textbook is dedicated to dark matter and dark energy. The primary reason for such "unfairness" to dark matter and dark energy is that we simply do not know much about these mysterious constituents of our Universe (but it would be nice if physics textbooks would emphasize just how much we do not know about the physical world).

Astronomical observations indicate that dark matter, which is an invisible form of matter that scientists have not yet directly observed, makes up about 25 percent of the Universe, and dark energy, a mysterious phenomenon or force that supposedly repels gravity, makes about 70 percent of the Universe. When we say that dark matter is invisible, we really mean that scientists have not detected any electromagnetic radiation (radio waves, microwaves, ultraviolet radiation, visible light, infrared radiation, X-rays or gamma rays) coming from dark matter.

Scientists hypothesize that dark matter exists because there is something invisible out there that has gravitational influence on galaxies and clusters of galaxies. For example, in any spiral galaxy, stars orbit the center of their galaxy. The laws of physics of ordinary matter tell us that stars located at the outskirts of a spiral galaxy's disk should travel slower than stars located closer to the center of the galaxy. However, observations reveal that stars orbit the center of any spiral galaxy at about the same speed, a speed that does not depend on their distance from the center of the galaxy. This "odd" motion can be explained by the presence of a halo of dark matter surrounding the galaxy.

When astronomers study other spiral galaxies and clusters of galaxies, they compare their gravitational mass and luminous mass. They determine the gravitational mass of a galaxy by studying how fast stars and clouds of gas and dust spin around the center of the galaxy. They determine the luminous mass of the galaxy by measuring the electromagnetic radiation emitted by all the matter in the galaxy (the luminous mass in galaxies can be in stars and interstellar gas and dust). So far, astronomers have determined that the amount of total matter in the galaxies exceeds the amount of matter that emits detectable electromagnetic radiation. That is, there is more undetectable ("invisible") matter in the galaxies than detectable matter.

Astronomers also use the idea of dark matter to explain the optical distortions of the images of distant galaxies. Such distortions can be caused by the clouds of dark matter located between the Earth and those galaxies. Dark matter might also explain the formation of galaxies and clusters of galaxies.

So far, all the information about dark matter comes from its gravitational effects on luminous (visible) matter. Because scientists have not had a chance to detect dark matter directly and study it in the lab, they can only speculate about its nature. Some theories propose that dark matter might consist of exotic particles, such as WIMPs (weakly interacting massive particles), that can exert a gravitational pull on ordinary matter, yet their interaction with ordinary matter or light is extremely weak.

Some scientists disagree with this idea. They propose that there is no dark matter, but that gravity itself behaves differently on a large scale. They theorize that when it comes to great cosmic distances as large as the size of galaxies, gravity obeys laws that are different from those governing gravity on our planet and within the Solar System.

The scientists who propose that dark matter is made up of exotic particles point out that there should be a large amount of dark matter in our Milky Way Galaxy (about 80 or even 90 percent of the mass of our galaxy could come from dark matter!). As the Solar System orbits the center of the galaxy, it might be traveling through a dense sea of dark matter particles, with many thousands or millions of dark matter particles filling each cubic meter of space!

This means that your body might be "bombarded" by these particle right now, while you are reading this book. Most of them would pass through your body, avoiding any interaction with the ordinary atoms that you are made of, but some of them might collide with atoms in your body.

A human body is basically an object made mostly of hydrogen, oxygen, carbon and nitrogen. Katherine Freese, a physics professor from the University of Michigan in Ann Arbor, and Christopher Savage from Stockholm University in Sweden, estimate that dark matter particles can collide with the nuclei of hydrogen and oxygen inside your body about 30 times per year. They admit though that the number of such collisions might be much greater, up to 100,000 times per year.

When it comes to highly energetic cosmic rays made of ordinary matter, we know that they can damage human body cells and cause cancers. Keep in mind that only an extremely small number of cosmic rays can be so harmful though. On the other hand, cosmic rays can also contribute to the evolution of life on our planet by changing human cells in some beneficial way.

If cosmic rays of ordinary matter can harm us, can dark matter particles bombarding our bodies be also unsafe? Freese and Savage admit that interactions between the cells of a human body and the dark matter particles can be a source of radiation in the human body. But they are rather optimistic about it. Their calculations indicate that the annual exposure to that radiation is negligible when compared to the radiation that comes from other natural sources, such as radon and cosmic rays, making the collisions with dark matter particles relatively harmless to us.

Nevertheless, their calculations are based on unconfirmed assumptions about the number, energy and properties of dark matter particles. Besides, we do not know for sure if dark matter is really made of particles or if it really exists. After all, dark matter might be a large-scale gravitational effect.

However, remember that dark matter is not the only invisible player in our Universe. There is also dark energy. The existence of dark energy was proposed in the 1990s when two independent groups of astronomers discovered that the expansion of the Universe is accelerating.

Scientists were not ready for that discovery at all! Until the discovery of the accelerated expansion, astrophysics relied on the assumption that there are four fundamental forces in our Universe that define four fundamental interactions: gravitational, electromagnetic, strong nuclear and weak nuclear.

Scientists expected that the gravitational force could slow down the expansion of the Universe.

On the contrary, it turned out that the expansion was not slowing down. It was speeding up! Scientists agreed that another type of fundamental repulsive force that possibly comes from space itself could explain such strange behavior. Scientists describe this mysterious phenomenon as *dark energy*.

How much do we know about it? As Inspector Jacques Clouseau from *The Pink Panther 2* movie said, "Let me bring you up to speed... We know nothing. You are now up to speed." Well, let us bring you up to speed… While various theories of dark energy are being proposed, scientists don't have any real knowledge of the nature of dark energy.

More than 20 centuries ago, the Greek philosopher Aristotle proposed that all the cosmic objects beyond the realm of our planet were made of the fifth element, called the ether or quintessence. Ironically, two thousand or so years later, scientists have to think of some kind of fifth force in order to explain the accelerating expansion of the Universe.

Many scientists found it reassuring to remember that Einstein's equations of general relativity originally included a special cosmological constant. In 1917, when Einstein worked on his theory general relativity, he did not know that the Universe was expanding. Because his equations didn't work for a static Universe, Einstein added to them a hypothetical repulsive force. He described it as the "cosmological constant." The constant was supposed to balance large-scale gravitational effects and keep the Universe static.

In the 1920s, Edwin Hubble, an American astronomer, used Cepheid variable stars, a special class of pulsating stars, as a "standard candle" to measure distances to galaxies. (Cepheid variable stars are the stars whose regular brightness variations are linked directly to their intrinsic luminosities so that their observed luminosities can be used by astronomers to estimate distances.) When his observations of distant galaxies proved that the Universe was expanding, Einstein reconsidered and withdrew his idea of the repulsive force. He called the cosmological constant his "biggest blunder."

Scientists brought Einstein's cosmological constant back to life in order to explain the accelerating expansion of the Universe. Today, one theory of dark energy proposes that dark energy represents that cosmological constant. The theory further speculates that dark energy has been steady and constant throughout the history of our Universe and will remain that way in the future.

Other theories say that dark energy is a manifestation of a new, fundamental force in our

Universe that only becomes noticeable when the expanding Universe reaches a certain size. The force might be temporary, making the Universe accelerate for a few billions of years and then disappearing

Let us be frank here. Even though adding terms to Einstein's equations can give a mathematical solution that is a closer fit to the observed acceleration of the expansion of the Universe, it doesn't explain the nature of the repulsive force. The equations tell us *how*, but they do not tell us *why*.

To explain dark energy, scientists propose another hypothetical feature, a sea of quantum zero point energy that makes space repel itself.

In any event, we know with certainty that the Universe is expanding faster and faster. In this expansion, something that we call dark energy stretches space itself. Can dark energy affect our lives on the Earth? In order to answer this question, let's take a look at some numbers. Scientists propose that dark energy has to be distributed evenly through the Universe. They estimate that dark energy's density should be approximately 10^{-26} kilograms per each cubic meter of space. (To compare, we can recall that the mass of one proton is 1.67×10^{-27} kilogram.) This means that all the dark energy in the Solar System is the equivalent of the mass of a small asteroid.

If dark energy is distributed so thinly, then its effect is insignificant at a level of a planetary system. However, it becomes noticeable on the scale of vast distances. In addition to its impact on the expansion of the Universe, dark energy might affect the evolution of clusters of galaxies, galaxies and stars.

Interestingly, the impact of dark energy on the formation and evolution galaxies might also affect the formation of planets, including planets similar to the Earth. If dark energy did not exist, then today's galaxies would be more massive populated with old generations of stars. Consequently, spiral galaxies similar to our galaxy would be rather rare.

If dark energy were more "powerful" in the past, then low-mass galaxies would be more common. In addition, fewer stars would have formed. Almost all the elements that planets contain come from old exploding stars. If the rate of star formation were lower, then fewer planets would have formed in the Universe. If this scientific reasoning is correct, then we should be glad that dark energy did not "come on too strong."

When it comes to the very distant future, dark energy might change our Universe in significant ways. If the Universe continues to expand faster and faster, then the repulsive dark energy will overpower gravity. First, it will break clusters of galaxies into separate galaxies. Then it will tear galaxies into "pieces." Finally, the accelerating expansion will destroy our Solar System and other stars and planets. Finally, it will rip molecules and atoms apart.

Scientists call this scary scenario *the Big Rip*. The likelihood of such a scenario depends on how dark energy behaves in the future—if it remains the same, becomes weaker or even disappears. Today, scientists cannot predict the future changes in dark energy because they do not understand its nature. However, if the Big Rip occurs, it will happen a few tens of billions of years in the future. Therefore, we have time to figure out what it is really going on with our Universe.

Part VIII. Life beyond the Earth

The artist's concept portrays Kepler-186f, the first confirmed Earth-size planet that orbits its star in the habitable zone. Image Credit: NASA Ames/SETI Institute/JPL-Caltech.

8.1. Can aliens from other planets knock at your door?

Our Solar System has many interesting objects and phenomena to explore. However, extrasolar planets orbiting other stars in our Galaxy might harbor something even more interesting: extraterrestrial civilizations!

Astronomers have discovered more than 1,800 extrasolar planets so far. The first discoveries revealed extrasolar planets that are completely different from the planets of our Solar System. Some of these extrasolar planets are as massive as Jupiter or even more massive than Jupiter, and they orbit extremely close to their parent stars. Radiation from the stars that they orbit heat them to very high temperatures and even blow their atmospheres away. Some of these extrasolar planets follow very elongated orbits, experiencing drastic changes in their surface and atmospheric temperatures.

As scientists continue perfecting their planet-hunting tools and techniques, they begin discovering Earth-like planets. For example, NASA's Kepler Mission surveys our part of the Milky Way Galaxy, looking for Earth-size and smaller planets located in or close to the habitable zones of their stars. (Scientists define the habitable zone as the region of space around a star within which a planet with appropriate atmospheric pressure can support liquid water at its surface.) The Kepler Mission also aims to estimate how many stars in our galaxy might have such planets.

In 2014, astronomers who used NASA's Kepler Space Telescope discovered the first Earth-size planet orbiting a star in the "habitable zone." The planet is called Kepler-186f, and it is located about 500 light years (5,000 trillion kilometers or 3,000 trillion miles) away from the Earth. According to all the data that scientists collected using the Kepler Space Telescope, about 20 percent of sun-like stars in our galaxy may have at least one Earth-like habitable planet.

The more similar an extrasolar planet is to the Earth, the more likely it is to be inhabited by Earth-like life forms and, perhaps, even by intelligent creatures. We suggest that Earth-like planets might harbor extraterrestrial civilizations because we already know of one civilization existing in the Milky Way Galaxy—our own. If we do not think of our Solar System and ourselves as "special and unique," then we should admit that intelligent life might emerge on extrasolar planets with suitable conditions.

In 1961, Frank Drake, an American astronomer, came up with an equation that would allow for estimating the number of potential alien civilizations in the Milky Way Galaxy. He limited his equation to the Milky Way Galaxy because the distances between galaxies are really enormous and radio communication with aliens from other galaxies would be practically impossible. It would take millions of years for radio signals to travel between galaxies. Besides, a radio signal arriving from another galaxy would be extremely weak and difficult to detect.

Drake multiplied a few variables together:

$N = R^* \times f_p \times n_e \times f_l \times f_i \times f_c \times L$

In this equation:

- N is the variable that describes the number of alien civilizations in the Milky Way Galaxy:
- R^* is the rate of sun-like star formation in the Milky Way Galaxy (today, astronomers include in this variable not only sun-like stars, but also stars that are smaller and cooler than the Sun, because they might also be orbited by planets with life forms):
- f_p is the fraction of those stars that are orbited by planets:
- n_e is the number of planets, per star, that are capable of supporting life (today, astronomers include in this variable not only planets, but some other "cosmic bodies" such as moons orbiting planets. This is because we know that our Solar System has three moons that may be suited for life: Jupiter's

moon Europa and Saturn's moons Titan and Enceladus);

- f_l is the fraction of those planets on which life emerges and evolves;
- f_i is the fraction of those planets on which life evolves into intelligent civilizations;
- f_c is the fraction of those intelligent civilizations that develop technologies suitable for interstellar communication; and
- L is the average time those civilizations are detectable.

Of course, scientists cannot really calculate the number of alien civilizations using this equation because they do not have accurate values for all its variables. However, they can try to estimate their values. L, which is the very last variable in the Drake equation, is difficult to guess though. We do not really know for how long a technological civilization can exist.

In the course of more than 50 years, various scientists have used the Drake equation to estimate the number of civilizations that have the means for interstellar communication. Different scientists got very different answers because their answers depend on the value of the variables entered. The most optimistic estimate that follows from the Drake's equation is that there are a few billion civilizations in our galaxy. The most pessimistic estimate suggests less than a hundred civilizations.

Even if a few million technologically advanced civilizations reside in our Galaxy now, they are scattered at great distances over the Milky Way Galaxy. The nearest extraterrestrial civilizations might be located dozens or hundreds of light years away from us. (One light year equals about 9 trillion kilometers or 6 trillion miles.)

Can we travel to their worlds? Can they travel to the Earth? Today, we do not have the technology that could send humans to other stars. If we traveled to Proxima Centauri, the star nearest to the Sun, and we were moving in space as fast as the *Voyager* spacecraft, then our trip would take about 80,000 years!

Theoretical physicists propose that there might be some space travel shortcuts such as wormholes. Scientists envision a wormhole as a tunnel that connects two different locations in the Universe, which are separated by huge distances in regular space. The wormhole provides a considerably shorter passageway between those locations. You should keep in mind that real wormholes have never been discovered. They exists only theoretically, as a mathematical solution to Einstein's equations.

The idea of wormholes is very popular among science fiction authors. Science fiction books and movies frequently "use" wormholes for space travel. However, the situation is different when it comes to the real space travel. According to theory, wormholes are very unstable. They would actually collapse if anything or anybody entered them. Besides, we do not know how to create wormholes, and we have not found any existing wormhole so far.

But that's not all! Even if you find or create a wormhole and travel through it, you might arrive in the wrong time epoch because the wormhole might connect two points in different space-times. Alternatively, the wormhole could actually send you to a completely different universe! It looks like wormhole travelers would have to travel at their own risk, and scientists who "design" wormholes would not be held responsible for any missed destination during the "wormhole" journey.

So far, we have not met any aliens who would have already used wormholes or other means of interstellar travel to get to our planet. There are various explanations for why alien space travelers have not knocked at our doors yet. For example, there might be a limited number of technologically advanced civilizations capable of interstellar travel. It is also possible that wormholes or any other space "shortcuts" do not exist after all, or that there are no wormholes that would bring aliens to the Earth.

Alternatively, interstellar travelers might not be visiting us because they find our civilization uninteresting or unacceptably aggressive. Another possibility is that technologically advanced civilizations might be short-lived, and many of them are already gone or have not arisen yet. Or, could it be that alien space travelers are already here on the Earth, but they prefer to stay unknown?

Hopefully, we'll find answers to these questions. Meanwhile, even if extrasolar planets do not "send" any aliens to the Earth, they do affect our civilization. This is because the very idea of the existence of intelligent creatures inhabiting other planetary systems inspires our scientists and engineers to seek ways of communicating with them or even travelling to their planets.

And, where there's a will, there's a way.

To Our Readers

Dear Reader!

Thank you for purchasing and reading this book.

You can find more information about science at the *Science Today and Tomorrow* website:

www.sciencetodayandtomorrow.com

To contact the author of this book, please visit the *Science Today and Tomorrow* website. You are welcome to send us your questions and comments.

References

1.1. How do shooting stars change our planet?

Chapman, Clark R., Peter T. Bobrowsky, ed., Hans Rickman, ed. 2007. Comet/Asteroid Impacts and Human Society: An Interdisciplinary Approach. Springer

Dalaia, Tarun K., Gregory E. Ravizzaa, B. Peucker-Ehrenbrinkb. "The Late Eocene 187Os / 188Os excursion: Chemostratigraphy, cosmic dust flux and the Early Oligocene glaciation." Earth and Planetary Science Letters, Volume 241, Issues 3–4 (January 2006): 477–492

Murad, Edmond, ed., Iwan P. Williams, ed. 2002. Meteors in the Earth's Atmosphere: Meteoroids and Cosmic Dust and their Interactions with the Earth's Upper Atmosphere. Cambridge University Press

1.2. How do meteorites make our planet more "precious?"

Dai Z.R,. Bradley J.P., at al. "Possible in situ formation of meteoritic nanodiamonds in the early solar system." Nature 418 (2002): 157-159

Goresy, Ahmed El, Philippe Gillet, et al. "In situ discovery of shock-induced graphite-diamond phase transition in gneisses from the Ries Crater, Germany." American Mineralogist. Vol. 86 No. 5-6 (2001): 611-621

Lewis, Roy S., Tang Ming, et al. "Interstellar diamonds in meteorites." Nature No. 326 (March 1987): 160 - 162

Marvin, U. B. "The search for Antarctic meteorites." Sky and Telescope, vol. 62, (November 1981): p. 423-427.

McKee, Maggie. "Diamond drizzle forecast for Saturn and Jupiter: Lightning storms create carbon soot that might be compressed into diamonds as it falls through the atmosphere." Nature, (October 2013)

Russian Diamonds: Siberian Meteorite Crater Said To Hold Trillions Of Carats. Huffington Post, September 17, 2012 [available at http://www.huffingtonpost.com/2012/09/17/russian-diamonds-siberian-meteorite-crater-carats_n_1891691.html]

Scandolo, Sandro and Raymond Jeanloz. "The Centers of Planets: In laboratories and computers, shocked and squeezed matter turns metallic, coughs up diamonds and reveals Earth's white-hot center." American Scientist, (November–December 2003): 525-530

1.3. Can another "Chelyabinsk" meteorite strike your city or town?

Chapman, Clark R., Peter T. Bobrowsky, ed., Hans Rickman, ed. 2007. Comet/Asteroid Impacts and Human Society: An Interdisciplinary Approach. Springer

Popova, Olga P., Peter Jenniskens, et al. "Chelyabinsk Airburst, Damage Assessment, Meteorite Recovery, and Characterization." Science, 29 Vol. 342 no. 6162 (November 2013): 1069-1073.

Atkinson, Nancy. "Airburst Explained: NASA Addresses the Russian Meteor Explosion." Universe Today (February 2013)

Wall, Mike. "Russia Meteor's Size, Origins Pinned Down By Scientists." Huffington Post (February 2013)

1.4. Where does our water come from?

Drake, Michael J. "Origin of water in the terrestrial planets." Wiley Online Library, Meteoritics & Planetary Science (January 2010)

Francois Robert. "The Origin of Water on Earth." Science Vol. 293 no. 5532 (August 2001): 1056-1058

Hartogh, Paul et al. "Ocean-like water in the Jupiter-family comet 103P/Hartley 2." Nature Letters, Vol. 293 no. 5532 (2011): 1056-1058

Mumma, M. J. et al. "Organic Composition of C/1999 S4 (LINEAR): A Comet Formed Near Jupiter?" Science Vol. 292, No. 5520 (May 2001): 1334-1339

1.5. Can comets and meteorites make life contagious?

Andrews Bill. "Curiosity Rover May Have Carried Bacterial Life to Mars." Discover Magazine. (May 2014)

Davis, John. "Paleontologist Presents Origin of Life Theory." (2013) [available at http://today.ttu.edu/2013/10/paleontologist-presents-origin-of-life-theory/#sthash.Xj8nPucD.dpuf]

Mumma, M. J. et al. "Organic Composition of C/1999 S4 (LINEAR): A Comet Formed Near Jupiter?" Science Vol. 292, No. 5520 (May 2001): 1334-1339

Rhawn, Joseph, and Chandra Wickramasinghe. "Comets and Contagion: Evolution and Diseases From Space." Journal of Cosmology, Vol 7, (2010): 1750-1770

Sun Kwok. 2011. Organic Matter in the Universe. Wiley-VCH.

Taylor, A. D., W. J. Baggaley, D. I. Steel. "Discovery of interstellar dust entering the Earth's atmosphere." Nature 380, (March 1996): 323 - 325

2.2. Can the Moon make life on the Earth more complex?

Peters, E. Kirsten. 2012. The Whole Story of Climate: What Science Reveals About the Nature of Endless Change. Prometheus Books

Shiga, David. "Aliens don't need a moon like ours." New Scientist, Issue 2838 (November 2011)

3.1. Can the positions of planets in the sky determine our fate?

Carlson, S. "A Double Blind Test of Astrology." Nature, vol. 318, (December 1985): 419-421

Kelly, I. "Modern Astrology: A Critique." Psychological Reports, vol. 81, (1997): 1035-1038

Kelly, I." Why Astrology Doesn't Work." Psychological Reports, vol. 82, (1998): 527-531

Kurtz, P. and Fraknoi, A. "Scientific Tests of Astrology Do Not Support Its Claims." Skeptical Inquirer, (Spring 1985): 210-214

3.2. Can Mars and the Moon throw rocks at our planet?

Agee, C. B.; Wilson, N. Vet al. "Unique Meteorite from Early Amazonian Mars: Water-Rich Basaltic Breccia Northwest Africa 7034". Science. Vol. 339 no. 6121 (January 2013). 780-785

3.3. Is Jupiter guarding our planet?

Hammel, H.B. "The Spectacular Swan Song of Shoemaker–Levy 9." 185th AAS Meeting 26. American Astronomical Society. ((December 1994): 1425-1428.

Valsecchi, G. "'236 Years Ago...' in Near Earth Objects, Our Celestial Neighbors: Opportunity and Risk." Proceedings of the 236th Symposium of the International Astronomical Union, Cambridge University Press, 2006

3.4. What happens when planets refuse to go in circles?

Batygin, Konstantin and Gregory Laughlin. "On the Dynamical Stability of the Solar System." The Astrophysical Journal, Volume 683 Number 2 (2008): 1207

Laskar, J. et al. "Existence of collisional trajectories of Mercury, Mars and Venus with the Earth." Nature, 459 (2009) 817-819.

3.5. Rogue planets: why are they not welcome in our Solar System?

Cartwright, Jon. "Homeless' Planets May Be Common in Our Galaxy." Science Now (May 2011)

Debes, John H.; Steinn Sigurðsson. "The Survival Rate of Ejected Terrestrial Planets with Moons." The Astrophysical Journal Letters 668 (October 2007): L167–L170

Technology Review. "Rogue Planets Could Harbor Life in Interstellar Space, Say Astrobiologists." Technology Review (February 2011)

4.1. Can we travel to the past without a time machine?

"Hubble telescope looks into past." [available at http://spaceplace.nasa.gov/review/dr-marc-technology/hubble.html"

4.2. Can the Sun fill our sky with clouds and send lightning to our planet?

Scott, C.J., R G Harrison, et al. "Evidence for solar wind modulation of lightning." IOP Science, Environmental Research Letters (2014)

Svensmark, H. "Influence of cosmic rays on Earth's climate." Physics Review Letters 81, (1998): 5027–5030.

4.3. Should we worry when the Sun turns the Northern and Southern lights on?

Coe, R. S.; Prévot, M.; Camps, P. "New evidence for extraordinarily rapid change of the geomagnetic field during a reversal." Nature 374 (April 1995): 687

Courtillot, Vincent. 1999. Evolutionary Catastrophes: the Science of Mass Extinctions. Cambridge University Press.

Merrill, Ronald T. 2010. Our magnetic Earth: the science of geomagnetism. The University of Chicago Press

Merrill, Ronald T., McElhinny, Michael W., McFadden, Phillip L. 1998. The magnetic field of the earth: paleomagnetism, the core, and the deep mantle. Academic Press.

4.4. How does the Sun turn off GPS systems and cause blackouts on our planet?

Balogh, Andre, Louis J. Lanzerotti, Steve T. Suess. 2008. The Heliosphere through the Solar Activity Cycle (Springer Praxis Books / Astronomy and Planetary Sciences). Springer

4.5. How does the Sun flood the city of Venice?

Barriopedro, David, Ricardo García-Herrera, et al. "A discussion of the links between solar variability and high-storm-surge events in Venice." Journal of Geophysical Research: Atmospheres, Vol. 115 Issue D13 (July 2010)

4.6. Can a solar grand minimum cause another ice age on our planet?

Hiroko Miyahara, et al. "Is the Sun Heading for Another Maunder Minimum? Precursors of the Grand Solar Minima." Journal of Cosmology, Vol 8 (2010): 1970-1982

Lockwood, Mike. "Solar change and climate: an update in the light of the current exceptional solar minimum." Proceedings of the Royal Society A 466 (2010): 303–329

4.7. What does the Faint Young Sun Paradox tell us about Earth climate theories?

Bengtsson, Lennart; Hammer, Claus U. 2004. Geosphere-Biosphere Interactions and Climate. Cambridge University Press

Goldblatt , Colin and Kevin J. Zahnle. "Faint young Sun paradox remains." Nature 474, E1 (June 2011)

Minton, David A., Renu Malhotra. "Assessing the Massive Young Sun Hypothesis to Solve the Warm Young Earth Puzzle." The Astrophysical Journal Volume 660 Number 2 (2007)

Peloza, Brian. "Purdue professor presents hypothesis for Faint Young Sun Paradox." Purdue University News Service [available at http://www.purdue.edu/newsroom/research/2012/120530MintonMigrating.html]

5.1. Can a supernova explosion destroy our Solar System?

Battersby, Stephen. "Earth odyssey: retracing our planet's turbulent galactic journey." New Scientist, Volume 212, Issue 2841 (December 2011): 42–45

Ratcliffe, Martin. 2007. State of the Universe 2007: New Images, Discoveries, and Events (Springer Praxis Books / Popular Astronomy). Springer

Stevenson, David S. 2014. Extreme Explosions: Supernovae, Hypernovae, Magnetars, and Other Unusual Cosmic Blasts. Springer

5.2. Exploding stars: are they good or bad for us?

Stevenson, David S. 2014. Extreme Explosions: Supernovae, Hypernovae, Magnetars, and Other Unusual Cosmic Blasts. Springer

Svensmark, Henrik. "Evidence of nearby supernovae affecting life on Earth." Oxford Journals, Science & Mathematics, Monthly Notices of the Royal Astronomical Society,Volume 423, Issue 2 (2012): 1234-1253.

6.1. Can our galaxy cause havoc on the Earth?

Filipovi?, Miroslav D , J. Horner, Evan J. Crawford, N. F. H. Tothill. "Mass Extinction And The Structure Of The Milky Way." Serbian Astronomical Journal (September 2013)

Matson, John. "Gamma-Ray Burst Fingered for Carbon 14 Spike in A.D." Scientific American (January 2013)

Minard, Anne. "Gamma-Ray Burst Caused Mass Extinction?" National Geographic News. (April 2009)

Miyake, Fusa, Kentaro Nagaya, et al. "A signature of cosmic-ray increase in ad 774–775 from tree rings in Japan." Nature 486, (June 2012): 240–242

Pavlov, Alexander A. et al. "Passing through a giant molecular cloud: "Snowball" glaciations produced by interstellar dust." Geophysical Research Letters Vol. 32 Issue 3 (2005)

Waller, William H. 2013 The Milky Way: An Insider's Guide. Princeton University Press

7.1. What can dark matter and dark energy do to us?

Freese, Katherine. 2014. The Cosmic Cocktail: Three Parts Dark Matter. Princeton University Press

New Scientist. 2014 The Unknown Universe: New Scientist: The Collection. Reed Business Information Ltd

Yun Wang. 2010 Dark Energy. Wiley-VCH

8.1. Can aliens from other planets knock at your door?

Farmelo, Graham and Hans Christian von Baeyer. "It Must be Beautiful: Great Equations of Modern Science." American Journal of Physics 70 (2002): 975

NASA's Kepler Mission [available at

http://www.nasa.gov/mission_pages/kepler/ma

Made in the USA
Monee, IL
04 November 2026

67379694R00085